SUNFLOWER HOUSES

Garden Discoveries for Children of All Ages

SHARON LOVEJOY

INTERWEAVE PRESS

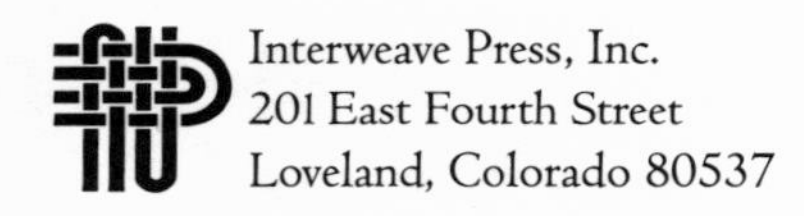

Interweave Press, Inc.
201 East Fourth Street
Loveland, Colorado 80537

Cover and interior design: Susan Wasinger, Signorella Graphic Arts

Library of Congress Cataloging-in-Publication Data:
Lovejoy, Sharon, 1945–
Sunflower houses : garden discoveries for children of all ages / Sharon Lovejoy.
p. cm.
ISBN 1-883010-00-4 : $14.95
1. Children's gardens. 2. Gardening. 3. Nature craft.
I. Title.
SB457.L83 1991
635--dc20 91-29880
CIP

First printing (perfect bound edition): 10M:494:OB

Printed in the United States of Americ

*Lots of people deserve bouquets of thanks for
their encouragement,
contributions and support.*

I would never have accomplished the thousands of hours of research, correspondence, travel, and tribulation had it not been for my husband Jeffrey Prostovich. His faith in me never wavered, he shouldered extra responsibilities in order to help me write, but most of all he taught me to believe in myself. My oldest friend, John Arnold, edited all of my work (a mammoth task in itself), and inspired me with his gentle love of flowers, birds, and bees. My dear friend Julie Whitmore searched for old poems, helped me locate some of the contributors, acted as a sounding board, and generally kept me buoyed up with her unflagging enthusiasm. Another dear friend, Betsy Williams, first mentioned my book to Linda Ligon who, I am happy to say, had the great, good courage to believe in me enough to publish this work. Thanks to Linda, and to Susan Wasinger of Signorella Graphic Arts for creative and sensitive design work.

And to my other friends, family, helpers, and gardeners: Grandmother, Abigail Baker Lovejoy, who started it all. Mother, Esther Lovejoy, who lovingly tended her little brick planters of pansies, begonias, and alyssum. Nonie, Augustine Elizabeth Clarke and Bopie, John R. Clarke, for their unending patience and love, letting me care for the "sprackets" and roses and their fascinating compost pile. Dear, sweet Gram, Gladys Marie McKinstry, who introduced me to her heavenly-blue morning glories and her old friend the Redbird. The unidentified lady who told me the story of her mother, Sunny Shoemaker, who created the yearly sunflower house (the inspiration for this whole book). Phyllis Shaudys; Portia Meares; The Hogue family— Emily, Tyler, Janna, Jack and Jane— of The Prairie Pedlar, Odebolt, Iowa; Christine Nybak Hill; Rhonda Schooley; Susan

DEDICATED TO

MY SON, NOAH

WM. ARNOLD,

WHO

REAWAKENED

IN ME THE JOY

AND PROMISE

OF LIFE

Jones Sprengnether; Gretel Hanna Barnitz; Millicent Truax Heath; Millie Baker Stanley; Thomas Stanley; Georgie Van de Kamp; Margaret Harper; Carol Bateman; Paul Mrozinski; Jan Blum of Seeds Blum, Boise, Idaho; Marjorie de Lyser; Patricia Kinser Reppert; Marjorie McClure Carnes; Margaret Crittenden Sparks; Chuck Molinari; Beth Benjamin; Ruth and Debbie Saltzman; Mary Allinson of Longwood Gardens; Sandy Tanck, Manager of the Youth Education Program at the Minnesota Landscape Arboretum; Catherine Eberbach; Brooklyn Botanic Gardens; Sheldon Cheney, National Agriculture Library Reference Librarian; Dorothy Fitzcharles Weber; Gina Douglas, Librarian and Archivist of The Linnean Society of London; Dolores Jackson; Dorothy Greeman Peterson; "Mammy"; Sarah and MaryBeth Monger; Judy and John Lewis; Betty McKinstry; Millie Heckman Huffaker; Freida, Mary, and Nikolai Reimer; Ricky Beauclaire; Carolyn Germain; Kim Cory; Roberta Baker; Diana Freeman; Ashley Leal; Mary Margaret Klug; Emily Rose Klug; Tracy Stevens; Stacy Willis; Dolin O'Shea; David, Julee, and Summer Krause; Josh, Jason, and Jonathan Prather; Jordan Marie Arnold; Robert Ball; my plant Nanny Jeanne Walker; Barbara Rogers of Herbitage Farm in Richmond, New Hampshire; John and Christine Tanner of Home Grown Herbs, San Luis Obispo, California; Frannie Norton, a researcher and exhibit aide at the Victorian Museum in Wheaton, Illinois; Phyllis Whitmore; Monica Zuvic (who kept Heart's Ease Gardens happy while I worked on this book); and last, but certainly not least, my pal Jane Dicus, who in the cold of autumn in Winston-Salem, went out searching of sassafras leaves for me. Thank you all and faretheewell.

Table of Contents

A note from the author

My first memories of home are of a tiny redwood cottage tucked into a vale in my Grandmother Lovejoy's garden. The vine-covered house was surrounded on all sides by old apricot and peach trees, and the lawn was carpeted with carnations. Raspberries and boysenberries grew along a wall, providing a secret nesting place for the night-singing mockingbirds.

My favorite haven in the garden was my playhouse—a pair of ancient guava trees that formed a huge, light-pierced tent with branches that swept the ground.

The pathway that led from Grandmother's house to mine was flanked with hollyhocks as tall as trees. Giant bumblebees nestled into the pink recesses of the blooms, while darting hummingbirds danced from petal to petal looking for unoccupied flowers.

The main spirit of our garden was a gigantic, mottled old sycamore tree with limbs so strong and comforting I would often curl up on the lower one to read or to watch for figures in the clouds. Grandmother's swing sat under the sycamore, and on sunny afternoons—with a gentle wind stirring the papery leaves—we would sip cream teas and eat sugar and

Grandmother Abigail Baker Lovejoy, who started it all...

In Grandmother's garden the hollyhocks

Row upon row lifted wreathed stalks

With bloom of purple, of pearly white,

Of close-frilled yellow, of crimson bright.

In Grandmother's garden the roses red

Grew in a long, straight garden bed,

By yellow roses with small, close leaves;

And yuccas—we called them Adams and Eves!

Threaded with fringes of fairy weaves;

By marigolds in velvet browns,

And heart's-ease in their splendid gowns;

Primrose, waiting the twilight hours.

Touch-me-nots, and gilliflowers.

Was it October, or June, or May

Grandmother's garden was always gay.

Sara Andrew Shafer

cinnamon sandwiches, and Grandmother would teach me about the flowers, trees, and animals of her garden.

On the day my Grandmother Lovejoy died, I ran to the shelter of my guava tree playhouse and closed myself inside for hours. I could hear the mocking-bird's young in the wall of berries, the wind rustling through our sycamore leaves, and the humming of the bumblebees working in the hollyhocks. I couldn't understand how the person who had given me this life could have gone, leaving these smaller things behind, unchanged.

What I have learned through the ensuing decades is that my Grandmother Lovejoy lives on. Her stories and teachings have enriched my life and the life of my son Noah for years. Now I pass this treasure on to you, and hope that you in turn will share the joys with the children in your life. Gentle lessons are waiting to be taught—and you, my friend, are the one to teach them.

Sharon Lovejoy
Cambria-Pines-by-The-Sea

Introduction

For the past dozen years, I have earned my living by growing, selling, and teaching about herbs and flowers. My passion for plants started when I was a youngster and was lovingly taught about gardens by my Grandmother Lovejoy—a teacher, naturalist, and one of the best gardeners ever.

My earliest and fondest memories are of endless days of summer play: hollyhock dolls under the peach tree, necklaces of rosebuds, wreaths of clover, daisy chains, and other simple pastimes Grandmother shared with me.

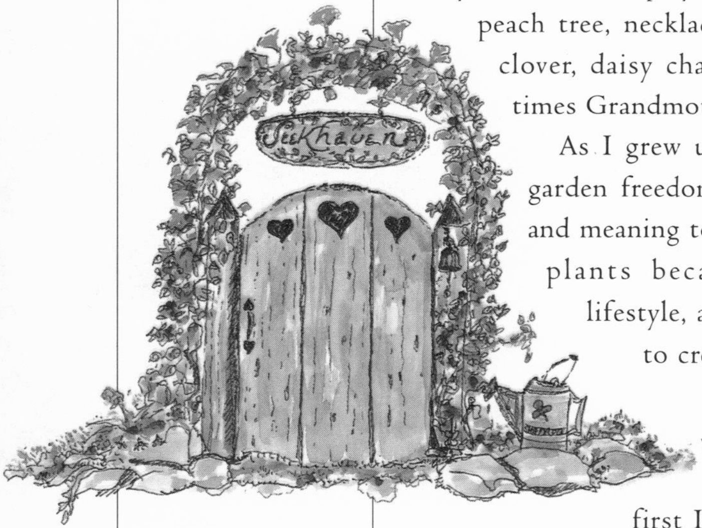

As I grew up, I realized that my early garden freedom had somehow given form and meaning to my whole life. My love for plants became the nucleus of my lifestyle, and I chose to devote myself to creating a beautiful community garden and to teaching about gardens and plants to people of all ages. At first I wanted to emphasize classes for children, but found, to my amazement, that adults became as animated and curious as the children. One of my oldest and most delightful students said, "I wanted to come and learn about all this before I got too old. I'm just 96 now."

I started thinking about what a garden had done for me, and to wonder about other's experiences in gardens. And so I began questioning people. "Can you remember any garden games or things you learned about plants and flowers when you were a child?" I would ask. Most times I could immediately tell if the person I was questioning had a memory to share. Generally there would be a start of recognition, a quick smile, and a nod, "Yes, I sure do, let me tell you about making trumpet vine dolls, acorn tops, and walnut sailboats."

All summer long we made belts, crowns, garlands – necklaces of tiny, perfect, pink rosebuds...

From my 7th summer

GENTLE LESSONS ARE WAITING TO BE TAUGHT

In 1983, I ran ads in *The Business of Herbs* and *Pot Pourri from Herbal Acres.* I asked people to share stories with me about their childhood experiences in gardens or with flowers. I received many responses to the ads, and I began interviewing people and collecting historical materials, garden plans, poetry, riddles, garden lore, and flower crafts. A true gleaning from hundreds of childhood memories!

In 1986, I started sketching some ideas for a very personal garden book that would incorporate my drawings of flower and plant projects, poetry, history, and first-person stories. That year my husband Jeff bought me an Apple Computer (which I promptly named Sarah Orne Jewett after one of my favorite authors) and said, "Get on with it, Sharon! Write your book." So I got on with it!

An author's note in an old book I found in Castine, Maine, a few years ago sums it up for me. "I can never repay the hollyhock the debt of gratitude I owe for the happy hours it furnished to me in my childhood." I feel that way, too, but perhaps my book will be a beginning.

Home

I WANT TO HAVE A LITTLE HOUSE
WITH SUNLIGHT ON THE FLOOR,

A CHIMNEY WITH A ROSY HEARTH,
AND LILACS BY THE DOOR;

WITH WINDOWS LOOKING EAST AND WEST,
AND A CROOKED APPLE TREE,

AND ROOM BESIDE THE GARDEN FENCE
FOR HOLLYHOCKS TO BE!

NANCY BYRD TURNER

A Child's Garden

Working and teaching in the gardens at Heart's Ease, I have always tried to note which plants most attract children and why. Some plants act almost as magnets, drawing children back again and again.

What are the secret qualities that make certain plants favorites of children? Well, my friend Georgie (who has been an avid gardener for nearly 80 years!) says, "What children really need is to have a lot of things that grow fast." She's right. Kids want to see some sort of quick response to the work and care they've showered on their piece of land. Also important: Personality (faces in a pansy, sunflower, or snapdragon); fragrance (once thought to be the very soul of the flowers); texture (woolly lamb's ear); and color—riotous, vibrant color.

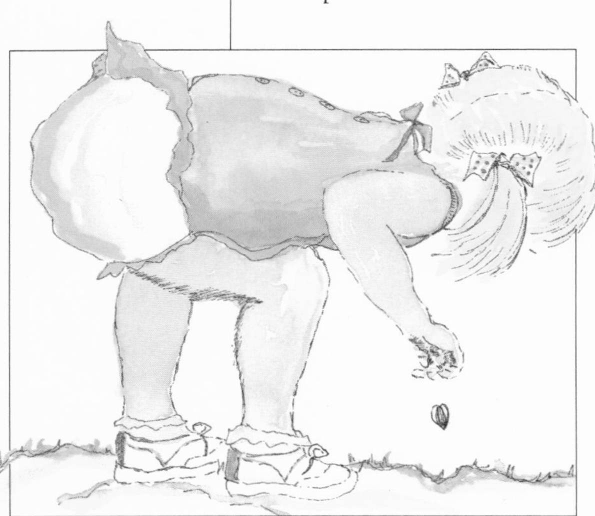

All outdoors
is warm and bright;

Robins sing with
all their might,

And the garden
seems to be

Just the place for
you and me!

John Bowman

Here are some of the plants I would choose for a children's garden, but remember not to limit yourself to just this list. The key to keeping children's sense of wonder and enjoyment alive is to let them make choices. Take the children to nurseries and let them search through the seed packets and flats of flowers for ones that twang their heartstrings. Allow them the freedom in their own gardens that they may not yet have in the rest of their lives.

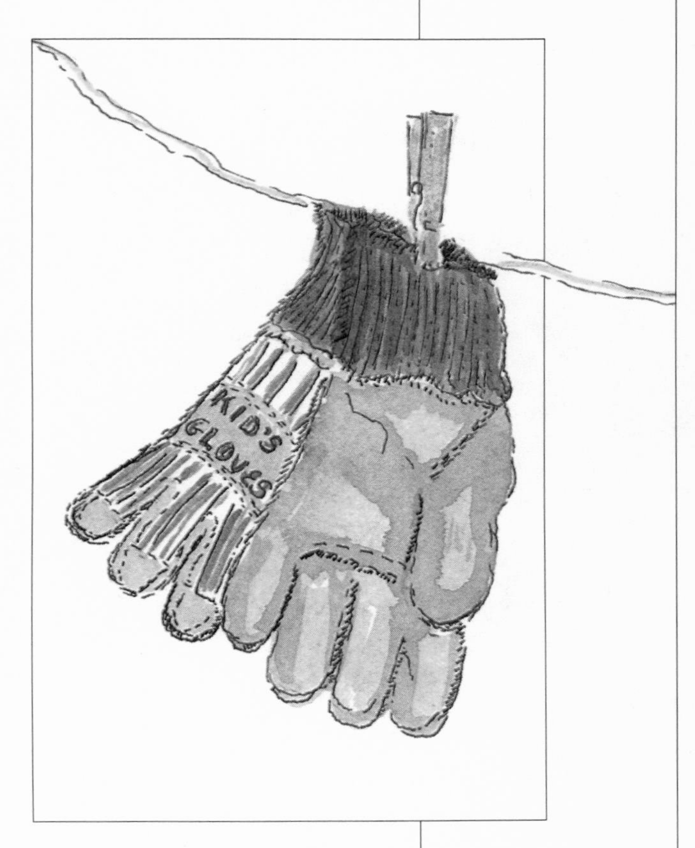

Perennials

BALLOON FLOWER *Platycodon grandiflorus*	Just plain fascinating.
BLEEDING HEART *Dicentra spectabilis*	For fairy gondolas or earrings; a perfect flower for pressing.
CHINESE LANTERN *Physalis alkekengi*	These brilliant orange globes look like real Chinese lanterns—perfect for a doll tea party. Exceptionally easy to grow and will self-sow prolifically.
COLUMBINE *Aquilegia spp.*	Also known as Little Doves or Granny's Bonnets, these are the fairy shoes of flower dolls.
DAY LILY *Hemerocallis spp.*	These are colorful and easy to grow, provided you have lots of sun. Children can gather the blossoms and buds and cook them. The flower blossoms make skirts for dolls.
DOLLAR PLANT, MONEY PLANT, OR HONESTY *Lunaria redeviva,* *or L. annua* (a biennial)	This makes play money, play dishes, gypsy jewelry. Easy to grow, and self-sows with abandon.

False Dragonhead
Physostegia virginiana

Kids can twist this into all sorts of funny positions and it will maintain the pose for hours (it is also referred to as Obedient Plant).

Lady's Mantle
Alchemilla spp.

Children love the dew diamond in the middle of the leaf (a gift from the fairies, no doubt).

Pink
Dianthus spp.

"Fresh pinks cast incense on the air, In fluttering garments fringed and rare." The scent remains in your memory forever. The petals are edible, and kids love to strew them in salads. Carnations (from the word coronet, and they can be made into coronets) are in this family, too. It's fun to string pinks together to make garlands and jewelry.

Poppy
Papaver orientale

They make excellent dancing dolls on their stems. Children use them as water ballerinas, and they are fun to watch opening and closing (sometimes we have helped open them). The seed pods are like tiny rattles or pepper shakers.

Stock
Matthiola incana

They have a sweet scent and are good in sand castles and in a moonlight or flower-dial garden. There are also annual stocks that are easy to start from seed.

PLANT RED FLOWERS FOR ME

The annuals we plant each Spring—

• • •

They perish in the Fall;

• • •

Biennials die the second year,

• • •

Perennials, not at all!

SUNFLOWER
Helianthus annuus

Sunflowers have true personalities, they attract birds and bees, they are the framework for our sunflower house, and the seeds are yummy. The mature flower heads can be harvested and used as birdfeeders. No children's garden can be considered complete without the whimsical presence of sunflowers.

I Meant To Do My Work Today

I meant to do my work today—
But a brown bird sang in the apple tree,
And a butterfly flitted across the field,
And all the leaves were calling me.
And the wind went sighing over the land,
Tossing the grasses to and fro,
And a rainbow held out its shining hand—
So what could I do but laugh and go?

. . . . *Richard Le Gallienne*

Annuals

CORNFLOWER, BACHELOR'S BUTTON
Centaurea cyanus

A good garland flower, and their rich blue and purple colors are jewels.

COSMOS
Cosmos bipinnatus

Brilliant and easy to grow, they attract lots of butterflies. My son Noah chose cosmos out of hundreds of plants in a nursery. They became his flower and were always planted outside our front door.

FOUR O'CLOCK
Mirabilis jalapa

These plants amaze me every day when they open at about 4 p.m., and at night they attract the wonderful, fascinating hummingbird moth.

CALENDULA
Calendula officinalis

The old-fashioned marigold. This is one of the plants that can be held under the chin to see if you like butter. Individual flower petals can be used in salads and in rice dishes. Sit your child down at a table and let him add petals to dishes as you cook. These flowers are easy to grow.

LOVE-IN-A-MIST
Nigella damascena

Its seed pods are wonderful—the tiny seeds shake out like pepper, and the pods can be used as teacups.

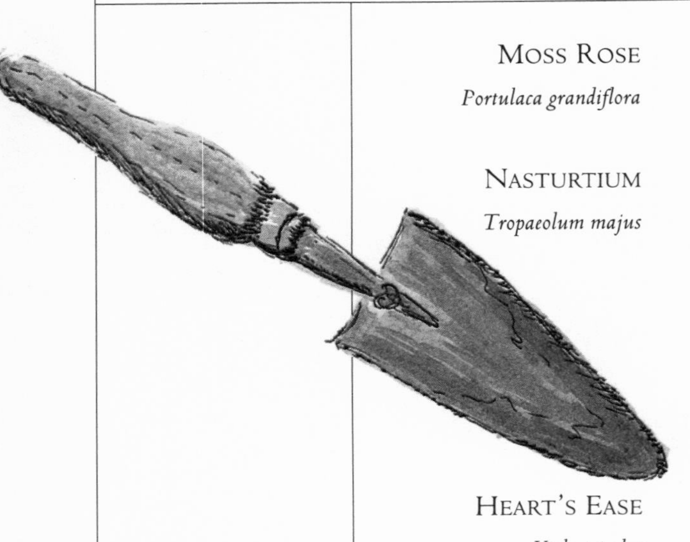

Moss Rose
Portulaca grandiflora

Brilliant and easy to grow, they belong in the flower-dial garden.

Nasturtium
Tropaeolum majus

The Elizabethans called them yellow lark's heels. These fiery charmers attract hummingbirds and children. With their peppery flavor, they can be used in salads or stuffed with cream cheese; children love to fill up the blooms and giggle as they eat them. The blossoms can be used in flower crowns and in leis and as hats on flower dolls.

Heart's Ease
Viola tricolor

Also called johnny-jump-ups. Who can resist those faces? Teach the children how to make heart's ease dolls.

Snapdragon
Antirrhinum majus

Along with hollyhocks, pansies, and sunflowers, this is one of the "personality" flowers. Snapdragons are great for hiding secret messages, making clip-on earrings, having snapdragon battles, or using like paper clips on each placecard at a child's birthday party.

Sweet Alyssum
Lobularia maritima

It spreads like a delicate carpet of snow, and children love the tiny, almost miniature bouquet blossoms. These flowers, like the tiny blossoms of thyme, fit perfectly into a child's fairy garden.

Sweet Pea *Lathyrus odoratus*

A little more difficult to cultivate, but worth the try just for the colors if not for the fantastic scent that children of all ages love!

Verbena *Verbena × hybrida*

Colorful and easy to grow, it attracts skippers and other butterflies.

Zinnia *Zinnia elegans*

Easy to grow and colorful, they also attract the skippers and other butterflies that bring so much life into a garden. A child-sized variety is fittingly called Thumbelina.

the • perfect • nursery!

Buy your eggs in the old-fashioned egg carton. • Tear off the lid, fill the egg holes with good potting soil. • Plant the seeds in each egg hole. • Water daily. • When the seedlings are about an inch high, gently tear or cut each egg compartment from the carton. • Each seedling may be planted still in its cardboard egg holder. • The cardboard will quickly disintegrate.

Biennials

Canterbury Bell
Campanula medium
Whole quadrilles of dolls with canterbury bell skirts can be formed on twigs and lined up for dancing.

English Daisy
Bellis perennis
Easy to grow, can be made into daisy chains, daisy grandmothers, garlands, and 'he-loves-me, he-loves-me-not' charms.

Forget-me-not
Myosotis sylvatica
Soft, green, mouse-ear leaves with sky blue eyes. These are grown easily in very shady spots and children love to use them in miniature bouquets.

Hollyhock
Alcea rosea
The all-purpose flower. They attract hummingbirds and bumblebees, they make great dolls, fairy goblets, and leis, and my favorite, firefly lanterns! The seed pods make good play money, too.

Sweet William
Dianthus barbatus
An old-fashioned flower that is extremely easy to grow and can even take a little neglect; it loves to be picked for bouquets or all kinds of flower projects.

Herbs

GARDEN CRESS
Lepidium sativum

A fast growing herb that gives children a feeling of instant success. Have the child write his or her name in cress seed, and watch it grow.

FENNEL
Foeniculum vulgare

A must for the back wall in your garden. This hardy plant provides food for the caterpillars of swallowtail butterflies. The caterpillars are incredibly beautiful and fascinating and the kids will watch the brilliantly mosaicked "pillars" eating their way from one branch to another. Fennel is a nibbler plant. Kids love to pick the tender, dark new growth and chew it like gum. It makes a soothing tea for upset stomachs, but best of all, it helps bring more butterflies into our world.

LEMON VERBENA
Aloysia triphylla

When children visit my garden, they always return again and again to the lemon verbena. When they leave, their pockets are usually stuffed full of the aromatic lemon-scented leaves. They know they can dry them for potpourri or use them in tea or just keep them in their pockets and pull them out for sweet smelling during the day.

My grass is green

My sky is blue—

I sing a song

of spring for you!

Mint
Mentha spp.

Easy to propagate from cuttings, its tangy, sweet leaves are great in teas, sachets, potpourris, and dream pillows. Kids just love to chew them. My favorite is the orange-bergamot variety.

Scented Geranium
Pelargonium spp.

Easy to grow from cuttings and tolerant of drought, they make a good houseplant for children. The leaves, smelling of rose, apple, lemon, mint, nutmeg, or dozens of other scents, are intriguing to kids as they hop from one to another and try to determine what they are smelling. Flowers can be used in cookies, teas, cakes, icings, syrups, fruit sorbets, fruit salads, and vinegars. Leaves are excellent in homemade potpourri mixes.

Woolly Lamb's Ear
Stachys byzantina

A touch-me, rub-me-on-your-cheek, and keep-me plant. Children spend lots of time at our woolly lamb's ear patch. I allow them to pick leaves and save them as bookmarks or to mail them off in letters. They are even more fascinated when I tell them that in olden days they were used as bandages!

Roses

Perfect,

doll-sized

teapots can

be made with

a rosehip,

stuck with a

thorn for

a spout.

Roses are not easy for children to grow or to play around, but they really are a must near a child's garden area. My favorite for ease of care, fragrance, and plump, beautiful rosehips, is the old-fashioned *Rosa rugosa*.

Children love to eat rose petals in rosy-cakes, they rub them on their lips as lipstick, or use them as fragile, tiny note paper for rose-notes to a special friend. Pea pod boats must have sails of rose petals or rose leaves! Perfect, doll-sized teapots can be made with a rosehip, stuck with a thorn for a spout. Necklaces can be strung with rosehips, and rosehip dolls can be made. You can brew up a cup of rosehip or rose petal tea for your tired little ones at the end of a long, hard day of play in the garden.

In California there is an Indian tribe that calls roses "Ska Pash Wee", which means, "Mean old lady, she sticks me!" Remember that rose thorns can be mean, so plant them out of the children's play area; but do plant them.

Vegetables

CARROTS
Daucus carota

The minis (called Little Fingers) are the best for children and they can easily be grown in a container. Kids love to pull the little finger-sized carrots and eat them raw. They're a hundred times better than the giant, woody, tasteless ones you get in a market. A swag of carrots can be hung on your door with ribbon to help greet the Easter bunny, or a bunny tussie-mussie can be made with carrots and radishes, surrounded by the herbs of your choice and beribboned like a floral tussie-mussie.

DECORATIVE GOURDS OF ALL TYPES
Cucurbita spp.

They can be used as bird houses, dippers, bowls, doll coaches, storage containers, baby rattles, false hen setting eggs, pretend dinosaur eggs, and Christmas tree ornaments. Young, tiny gourds can be made into gourd dolls. A variety called the spaghetti gourd or squash is a delicious, edible oddity as tasty as the pasta for which it was named.

EASTER EGG EGGPLANT
Solanum melongena

An aptly named ornamental edible that can be grown in containers. Fruit is satin-white and egg-shaped.

Fraises des bois, or Alpine Strawberries
Fragaria vesca

Not a vegetable, but these should have a place somewhere in your garden! The kids call them fairy berries at Heart's Ease and they head straight for the berry borders to fill up on the tasty little gems. The love children feel for these berries cannot be equated with their size!

Painted Lady Beans
Phaseolus coccineus var.

Heirloom British pole beans that love to clamber over trellises, arches, and teepees. This bean flowers prolifically and produces beautiful coral and white flowers with a sweet, light bean flavor—yes, the blooms are edible and can be used in salads, sandwiches, and as a great snack.

Pumpkins
Cucurbita pepo

Both the mini and giant varieties are great personality plants to have in a garden. Pumpkins can be scratched with a nail, and as they grow the etched name or message grows.

Radishes
Raphanus sativus

They are so easy to grow that they make children feel like master gardeners. Several seed companies now have a variety of radishes called "Easter Egg". These delicious round radishes are colored lavender, red, pink, and white. What a surprising treat for kids!

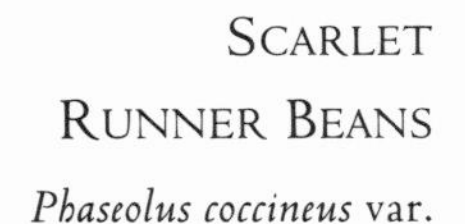

SCARLET RUNNER BEANS
Phaseolus coccineus var.

Fast growing, attracts hummingbirds, and makes excellent teepees. The red flowers are edible.

STRAWBERRY POPCORN AND MINIATURE INDIAN CORN

Zea mays var.

These are kid- sized. The strawberry popcorn produces a truly strawberry looking head of corn that can be popped (if you have the heart). The mini Indian corns are loved by children because of their charm and tiny size; they make wonderful miniature corn husk dolls. 'Pretty Pops' has confetti-colored kernels of red, blue, orange, black, yellow, and purple that show white when popped. The ears are a great Thanksgiving or autumn table decoration.

TOMATOES
Lycopersicum spp.

These can make children feel as if they are really accomplishing something. 'Currant' tomatoes are pea sized and would be perfect in a miniature vegetable garden. 'Yellow Marble' is, you guessed it, the size of marbles.

Vines

Cup and Saucer Vine *Cobaea scandens*	A ready-made tea set. Flowers open yellow-green and age to a rich purple.
Honeysuckle *Lonicera japonica*	This sometimes pesky vine has the sweet-tasting flower children love to suck. The vine attracts hummingbirds and the long, supple branches are easily woven into baskets and crowns. The flowers can be stuck inside each other, forming a continuous chain of blooms. Children wear the strings of blooms as crowns, belts, and necklaces.
Hops *Humulus lupulus*	Fast growing; so fast, in fact, that people often place bets on how much a vine will grow during the hot summer nights.
Moon flower *Ipomoea alba*	Attracts night moths, including the fascinating hummingbird moth; it is fun to watch the flower as it opens at night.
Morning Glory *Ipomoea purpurea*	Certain varieties attract hummingbirds, and the 'Heavenly Blue' is the roof of the sunflower playhouse (page 61).
Trumpet vine *Campsis radicans*	Attracts hummingbirds, makes great dolls and bubble blowers.

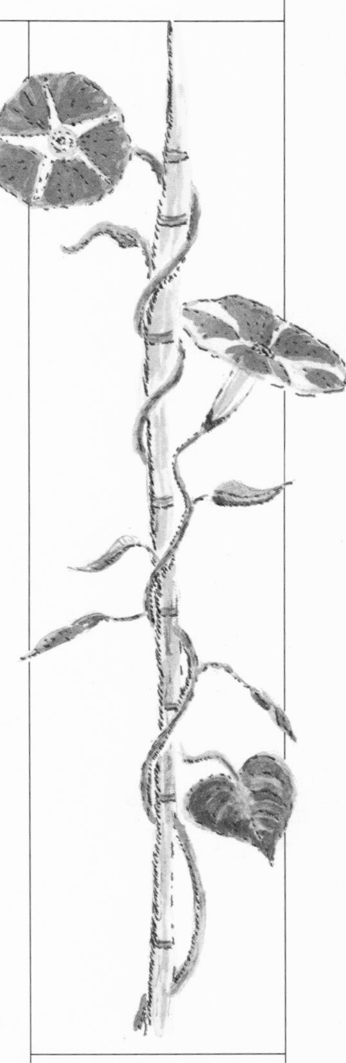

A Giant Garden

I like to spend my "free" time reading seed catalogues. Some great ones are available, and I think I get them all (my postman knows I do). One of my very favorites is the Seeds Blum catalogue from Jan Blum of Boise, Idaho. It is chock full of information, ideas, and inspiration. Thus sprouted the idea for Jack-In-The-Beanstalk's Giant garden.

If you have a rough-and-tumble child who is easily bored, tantalize him (or her) with a garden that could be the biggest, best, most unique one in town. Remember, this garden needs plenty of space for the plants to roam—and roam they will!

> We've laughed round the corn heap
> With hearts all in tune,
> Our chair a broad pumpkin—
> Our lantern the moon.
> Telling tales of the fairy who travelled like steam
> In a pumpkin shell coach
> With two rats for her team.
>
> *John Greenleaf Whittier*

A tomato called 'Delicious' is listed in the Guinness Book of World Records as the largest tomato ever grown—6 pounds and 8 ounces. Wow! A variety of pumpkin called the 'Cinderella' is large

Imagine Vegetables Like These:

• • • • •

Big Moon pumpkins (400 pounds)

• •

Oxheart tomatoes (4 pounds each)

• •

Scarlet Imperial Long carrots (3 feet long)

• •

Giant Perfection muskmelon (15-18 pounds)

• •

Grey Stripe sunflower (10 feet tall)

• •

Scarlet runner beans (12" pods on plants that can be 30 feet tall)

• •

Walla Walla onion
(1½ pounds)
••
Italian parsley (3 feet tall)
••
Lagenaria summer squash
(2-3 feet long)
••
Zwaan Jumbo cabbage
(20-35 pounds)
••
Aconcagua pepper
(12 inches)
••
Armenian cucumber
(3 feet long)
••
Crimson Long radish
(1 foot long)
••

enough for a person to ride in! The 'Cinderella' has reached nearly 500 pounds, another world's record.

Take a playful approach to designing your giant garden. Don't just plant plain rows—use trellises (how about old ladders?), teepees and arches. Children need to be able to play among the plants and to interact with them. Children need to feel the energy of the plants as they grow—in fact, it's something everybody needs, not just children.

THE SOIL
Teach
IS THE
children
PLANT'S
that:
DINNER!

"You can't just plant regular vegetables and expect them to become giants," Jan Blum says. The varieties mentioned here are genetically larger. There are some tricks to growing the biggest, but they are just tricks. Some people pick off all but one ear of corn, or one pumpkin, in order to allow all of the growth to go to one, special vegetable. Some people sow seed on piles of manure and compost and actually milk-feed the plants! But the most important thing for growing any kind of healthy plant is good soil!

The Girls

At my home I have a 4 feet x 8 feet worm box. Yes, you read it right, a worm box. We fondly refer to our worms as "the girls". I often ask for leftovers from restaurants for my girls, though this is sometimes puzzling to waitresses who ask, "You mean you are taking your soggy, left-over salad home to your daughters?"

My worm box is very simply constructed of common cinder blocks stacked three high, with one of the tiers of blocks below ground level (we didn't want any runaways). My husband cut two pieces of plywood which sit side by side on top for a lid. I use one side of the box one week and then switch over to the other.

I got my start of red earth worms from my gardening friend George Kryder, who feels that "the girls" really love and thrive on leftover bread and cake and cold pasta. I started with one bucket of girls

"Sunshine and water,

Love and good soil,

Weed it and feed it,

Harvest your toil."

and found that soon I was the mother of millions. The girls feast on my kitchen refuse and shredded clippings and within about a ten-day period, they are able to consume vast quantities of food, pass it through their digestive systems, and produce rich, thick, chocolaty castings, which are like gold for the plants.

My worm box is clean and odorless (I keep it covered because the robins and the raccoons consider it their private dessert tray). I water the girls often (they like their living quarters moist) and feed them daily, and once a week we take a pitchfork and turn them and remove about a wheelbarrow full of "gold" and girls for our garden. Our small box has been the parent to many other worm boxes in our town, and everyone is happy with this thrifty and organic solution to a sometimes overwhelming garbage problem. Then there is the satisfaction of working with the girls—so quiet and industrious.

A Floral Clock Garden

Young Joy ne'er

thought of counting

hours,

'till Care, one

summer's morning

Set up, among his

smiling flowers

A dial by way of

warning.

MURRAY

Years ago, I read a story about a floral "clock", called "The Garden of Hours". It was created by the great botanist Carl Linnaeus at Hammersby in northern Sweden. In trying to find information about the clock, I wrote to the Linnaean Society in London, England.

The society librarian and archivist kindly sent me a photocopy of the clock as given in Linnaeus' *Philosophia Botanica* of 1755. The clock, the archivist said, is only of academic interest to us because it was designed for latitude 60° north. Thus, it would not work accurately in the United States.

Linnaeus was obsessed with the idea of his flower clock, or "Watch of Flora", as he often referred to it. His watch was composed of forty-six flowers that opened and closed at predictable times through the hours of the day.

I kept mulling over the idea of a flower clock in a children's garden. What a great way to teach children about the characteristics of flowers. By observing opening and closing and rain and sleep behavior, a child could learn about the individuality and miracle of each plant.

I began collecting information on the opening and closing times of plants. The following information comes from my own field work and the writings of botanists of the early 1800s. From it, I designed a child's flower dial that grew first in my mind, then on paper. Finally, it flourished in the soil. Choose from among the plants I've listed those that will do best in your climate. Opening and closing times will vary greatly from one locale to another, so observe your plants carefully, and transplant as necessary!

First, pace off a circle, from a few to ten feet across. Outline your clock circle with rocks and then divide it into segments just like the face of a clock. Run a line of rocks through the center of each segment to separate the times into a.m. and p.m. Put a post two to three feet tall in the middle of the circle or a sundial on a pedestal for a nice focal point.

You can outline your clock garden with a suitable edging plant such as box, germander, or a small lavender such as 'Hidcote' or 'Munstead'. Choose something short and compact that will not shade the other plants.

A FLORAL SUN DIAL

How well the skilful gardener drew • Of flowers and herbs this dial new... • And as it works, the industrious bee • Computes its time as well as we! • How could such sweet and wholesome hours • Be reckoned but with herbs and flowers!

Andrew Marvell

'Twas a lovely thought

2:00 a.m.

The fragile, folded purple Convolvulus shakes out her silky skirts.

3:00 a.m.

The Egyptian Waterlily and Goatsbeard open slowly just as the Campions are calling it a day.

4:00 a.m.

Spiderwort greets the day just as the beautiful, sky-blue Flax opens its twinkling eyes.

5:00 a.m.

Chicory is also an early riser. In olden days, chicory was called Ragged Sailors or Miss-go-to-bed-at-noon. That name makes it easy to remember when it goes to sleep.

6:00 a.m.

Morning Glories and also Day Lily, Iceland Poppy, Hawkweed, and Cape Marigold bloom between 5:00 and 6:00. Dandelions awaken early, too, but you probably won't want to plant them in your garden!

to mark the hours,

7:00 a.m.

Madwort, African Marigold, St. Bernard Lily, White Water Lily (find an old tub and fill it with water for this one), and Fig Marigolds begin to open.

8:00 a.m.

We can look for the friendly greeting of the Scarlet Pimpernel and the Fringed Pinks. If it is a cloudy day, the Scarlet Pimpernel will not open. In olden days, it was called Poor-man's-weatherglass because it always closed before a storm.

9:00 a.m.

Marigolds, Tulips, Ice Plant, Pink Sandwort, Chickweed, Mallow, Moss Roses and Gazanias look sunward. Dandelion closes at 9 a.m., as do the Water Lilies.

10:00 a.m.

California Poppies and Golden Stars are fully open. Evening Primrose has changed from white to rose to bright pink as the limp, silken flowers close.

11:00 a.m.

Star-of-Bethlehem closes, as do the common Sowthistles lurking along the borders of the garden. Passion Flowers awaken and Sweet Peas embrace the day. What sleepyheads!

12:00 noon

The Chicory is already retiring and the Pinks begin dozing off as the Goatsbeard quietly folds away. The wild Daisies (or day's eyes) are opening.

As they floated in light away,

1:00 p.m.

There's a hush over the garden during the hour after midday, while open flowers continue to follow the sun, and later bloomers wait for the heat of noon to pass.

2:00 p.m.

Scarlet Pimpernel is getting sleepy and the weary Moonflower is preparing for bed time. Pinks join the Moonflower in slumber. Tulips and Daisies are at their fullest.

3:00 p.m.

Vesper Iris opens briefly, a lavender haze of small flowers. Field Marigolds, red Sand Spurry, Ice Plant, Hawkbit, Fig Marigold, and Pink Sandwort call it a day.

4:00 p.m.

Four-o'clocks, called the Marvel of Peru, rise at, you guessed it, four o'clock! Cape Marigold, Madwort and St. Bernard's Lily drift off to sleep.

5:00 p.m.

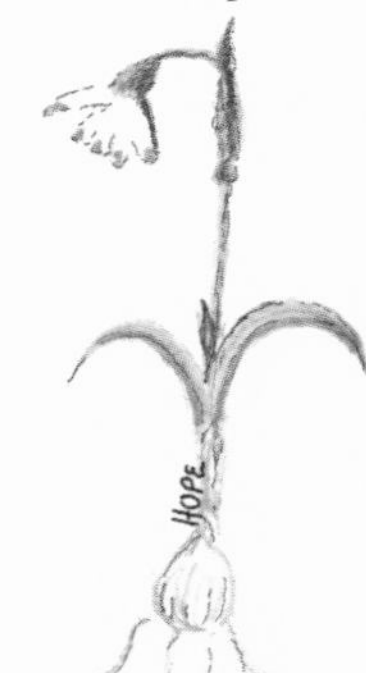

Evening Primrose, and Jimson Weed begin their debut at 5 p.m. I have read stories and poems about the opening of the Evening Primrose, but I have never been able to catch it in action as Keats must have when he wrote that he was "startled by the leap of buds into ripe flowers." I once heard a child refer to the luminous, fragile blooms as "fairy tents". Cat's Ear closes, and all of the Water Lilies are napping. Buttercup's glow is folded away for the evening.

By the opening and the folding flowers,

6:00 p.m.

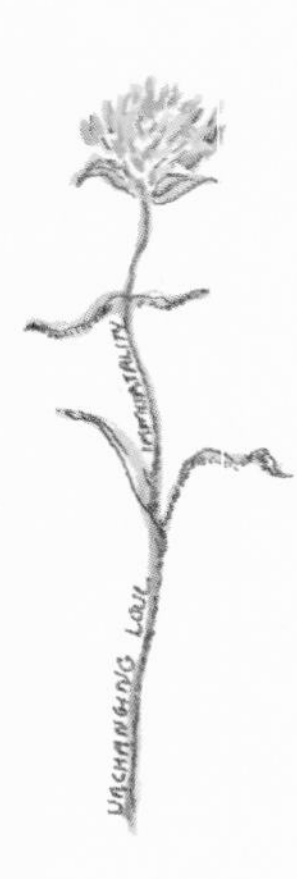

The Evening Primrose is in full glory. One species of Primrose, *Oenothera hookeri*, opens with "patens of bright gold", as Margaret Armstrong said, shining through the deepening shadows. Nottingham Catchfly scatters white stars across the garden floor, as Honeysuckle opens. As the sun sets for the day, the beautiful, fragrant, white Moon Flower with its ghostly bloom opens slowly and scents the garden with a haunting clove-like fragrance. Grow the elegant Moon Flower at the edge of your clock, perhaps on a low fence or trellis. In olden days, it was common for the whole family to sit on the porch waiting for the opening of this striking bloom. Quietly, they watched as the long white buds moved almost imperceptibly and then the great, white, starfaced flower opened slowly to release its fragrance and welcome the visits of the giant hummingbird moth (a favorite of every child!).

7:00 p.m.

The Iceland Poppy closes as the white Evening Campion awakens. This night-bloomer attracts many interesting night-flying moths. Sweet White Tobacco "...wakes and utters her fragrance In a garden sleeping" (Edna St. Vincent Millay).

8:00 p.m.

The tawny Day Lily slumbers as one of my favorites, the night-scented Stock, unfolds its small purple flowers. In olden days this inconspicuous plant was called the Melancholy Gilliflower because it looks so sad during daylight hours.

9:00 p.m.

Moon Flower should be fully opened by now, joined quietly by Sweet Rocket, often called Daughter-of-the-Evening because she only releases her scent after the sun is down. Postage Stamp Plant, so-called because of its fringed flower petals, awakens in the coolness of evening. When the flowers open, the air is permeated with the smell of almonds and vanilla.

10:00 p.m. to Midnight

Between 10 p.m. and midnight the strange pageant of the Night-Blooming Cereus begins and ends. This regal beauty is called Queen-of-the-Night—you'll soon understand why! I remember being allowed to stay up way past bedtime so that I could watch the Queen awakening. Brilliant Moss Rose prepares for sleep.

1:00 a.m.

Just as in the hour after noon, a hush falls over the garden.

Shall I sing of
happy hours
Numbered by opening
and closing flowers?

HARTLEY COLERIDGE

Remember that these hours will probably not be just right for your garden. Some of the sleeping and waking will be up to you to watch and record. You may wish to find other plants to tuck into your clock garden and you may need to move some of the suggested plants from one time area to another.

Flowers often take short naps during the day. You must go outside and watch quietly. Sometimes when the sun goes behind a cloud the scarlet pimpernel and poppies may tuck away for a bit. Some flowers close to protect their pollen from being washed off by an unexpected storm, so they may close many times during the day. Take time to really look at how the flowers sleep. Some bend their heads; others, like oxalis, turn into little tent or umbrella shapes. Look at the tight petals of the California poppy. It is called Dormidera, little sleepy head. Can you see why? So much is going on in the quiet, green world of plants!

YOU MUST

Flowers often

GO OUTSIDE

take short naps

AND WATCH

during the day.

QUIETLY.

A Good Word for the Bugs

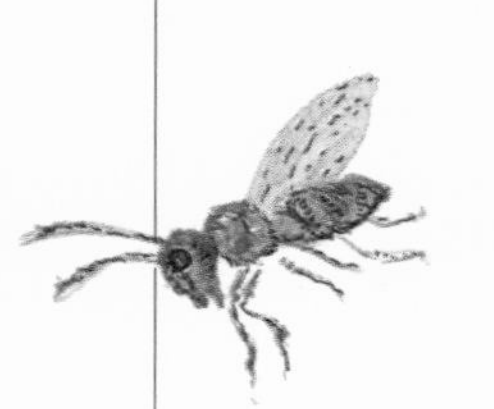

I think it is about time for someone to speak up for all of the good bugs out there! I am always startled when I have a group of children in our garden and hear one shout with fright at the approach of an innocent insect. I have to laugh and point out the size difference, asking, "How would you feel if some huge, lumbering thing came walking toward you and you were just a tiny spider?"

Children fear what they do not understand. A good case in point was the unreasonable terror I felt as a child in a garden full of dragonflies! As soon as the dragonfly was explained to me, the fear disappeared.

Take the time to introduce your child (and yourself) to the lady bird beetles, praying mantises, and other interesting and beneficial bugs who feed on the larvae and adults of destructive insects. Hosting a healthy population of these friendly critters can help make your garden a thriving, pesticide-free environment. You'll be happy they chose to call your garden home.

Emily's Kinder Garden

I'm always especially happy to receive my mail when I spy the handwriting of my friend Jane Hogue, The Prairie Pedlar, gardener extraordinaire, and mother extraordinaire! Her letters are always beautiful and inspirational and this one about her daughter Emily's kinder-garden was exceptionally so.

Emily's garden was begun the summer before she was old enough to go to school, and before she knew her ABCs. But she learned them quickly and in the most enchanting manner possible—she learned by planting and taking care of her own tiny garden filled with plants from A to Z.

Jane hoped that in caring for the garden Emily would learn not only her ABCs, but that she would learn to identify the flowers by association. "For example," Jane wrote, "C is for cockscomb that looks just like the comb of the rooster; D is for daisy—'He loves me, he loves me not'; L is for lamb's ear, which feels just like the real thing; R is for rose—we included a ring of mulch around the miniature rose bush, assuming the garden creatures might want to play 'Ring Around the Rosy'; and S is for sunflower—nature's birdfeeder."

A
ASTER

B
BACHELOR'S BUTTON

C
COCKSCOMB

D
DAISY

E
EMILIA

F
FORGET-ME-NOT

G
GLADIOLUS

H
HOLLYHOCK

I
IRIS

J
JOHNNY-JUMP-UP

K
KALE

L
LAMB'S-EAR

M
MARIGOLD

N
NASTURTIUM
O
OBEDIENT PLANT
P
PETUNIA
Q
QUEEN-ANNE'S-LACE
R
ROSE
S
SUNFLOWER
T
TANSY
U
URSINIA
V
VERBENA
W
WORMWOOD
X
XERANTHEMUM
Y
YARROW
Z
ZINNIA

Emily's garden was laid out in two tiers and snuggled up against the side of Jane's flower drying room. Forming a background to the garden was a friendly line of sunflowers. Each flower and letter of the alphabet was marked by a decorated slate displaying both the letter and the flower. What better way to learn your ABCs and flowers?

In Emily's garden, the letter U was represented by "Unions". Yes, you read right, Emily spells onions "unions", and if you say the word aloud, she is pretty close to right.

For U in my kinder-garden, I chose Ursinia, which is a South African plant in the aster family. Ursinia produces a bright orange daisy and blooms profusely from late summer to the first freeze.

Tyler's Barnyard Garden

COLT'S FOOT
•
BEE BALM
•
PUSSY WILLOW
•
CORNFLOWERS
•
COWSLIP
•
DOGWOOD
•
STRAWBERRY
•
GOOSEBERRY
•
CHICKWEED
•
GOOSE GRASS
•
HOG APPLE
•
PIG-STY DAISY
•
OX-EYE DAISY
•
MILK MAIDS

Tyler Hogue is a true Iowa farm boy. He loves to spend his daylight hours outdoors exploring and working in his own "Barnyard Garden".

Young boys love tools of all kinds, and Tyler is no exception. Fascinated by an old plow sitting out in the garden, he chose to adopt the plow area as his own.

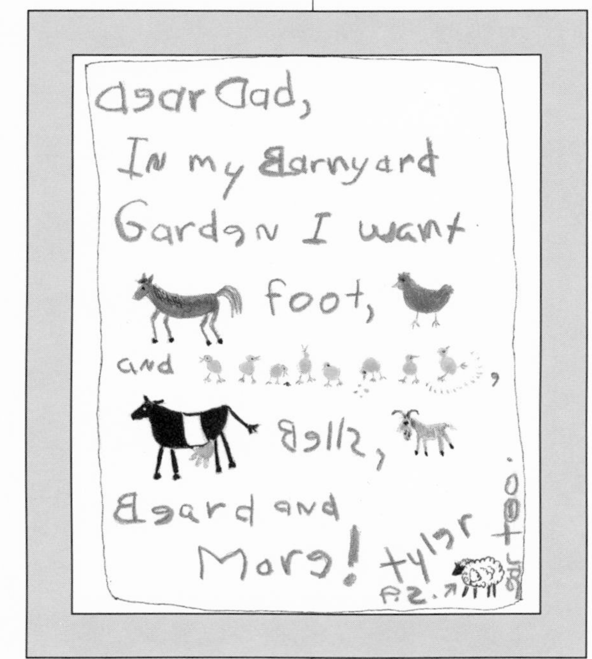

Tucked in and around the plow are the plant denizens of a farmyard. Hens-and-chicks, Lamb's Ear, Cowbells, Horseradish, Horehound, and Goats Beard vie for their share of sunshine. The plot is just the right size for a ten-year-old to tend. His mother, Jane, worries that another barnyard dweller might take over—Pigweed!

Perhaps you can add even more plants to the barnyard list.

Fish in a Bottle

How in the world did a huge zucchini get into such a snug place?

A woman visiting Heart's Ease told me that when her children were growing up she liked to surprise them with "different" gifts from her garden. In the spring she would let each child choose a large, empty bottle from her storage pantry. Then they would go out into the vegetable garden and head for the zucchinis and cucumbers where each child would "catch a fish".

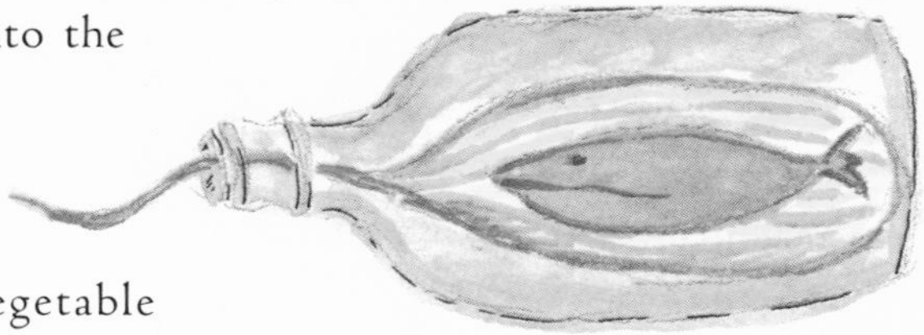

The children would each find a tiny, newly-formed cuke or zucchini and, with a pencil or nail, scratch a fish shape into its side. Gently, slowly, the children slipped each tiny, forming vegetable through the neck of a bottle and into the body.

Every day the children would go outside to check the progress of their fish until finally, before their very eyes, the tiny fish had filled the whole bottle and become a giant whale. Neighborhood children, eyes bulging in wonder, would turn the bottles over and over, trying to figure out how in the world a huge zucchini had gotten into such a snug place.

Mini Trough Garden

Jan Blum and I were talking about the Jack-in-the-Beanstalk Garden and she said, "If I were a young child, what would really thrill me would be the mini vegetables. In fact, I'm a grown-up and I am thrilled with them—the tiny currant tomatoes just send me over the edge."

In my "Tot's Garden" at Heart's Ease, I have made a miniature vegetable garden in an old English rock trough. The rectangular trough measures 24 inches by 18 inches and is 11 inches deep. It's knee high for the little knee-high gardeners who visit us.

A thick layer of crushed crockery and pebbles lines the bottom of the trough. A good, fast-draining planting soil (enriched with castings from my girls) fills the trough nearly to the top.

And in the trough are lots of wonderful mini-vegetables planted in tiny, symmetrical rows. I have miniature plant markers

Suggested Plants

'La Belle'—mini filet bean

'Little Ball'—mini baby beet

'Planet'—a small, ball-shaped carrot

Cornichons—European-style mini cucumbers for pickling

'Little Fingers'—a baby eggplant

'Little Gem Mini Romaine'—lettuce

'Rubens Dwarf Romaine'—lettuce

'Summer Baby Bibb'—lettuce

'Tom Thumb'—lettuce

'Early Aviv'—baby onions

made out of popsicle sticks. The markers are decorated with tiny drawings of the plants in each section.

If your heart runs towards mini-flower gardens, there are many varieties of plants to choose from. 'Little Sweetheart' sweet-peas are a natural for growing in pots or troughs. Many of the bulb catalogues now feature miniature daffodils, roses, and irises. My early spring borders are filled with tiny, purple grape hyacinth. Living up to their name, they emit a light, grape scent on a warm day. My borders host snowdrops, 'Tiny Rubies' dianthus, 'Peter Pan' saxifrage, and innumerable other minis that would be perfect for a pint-sized garden.

Let your children read through your seed catalogues. Take them to local gardens, parks and arboretums. Let your enthusiasm overflow and enjoy your child's awakening awareness of the endless possibilities of gardening—even in the smallest, most unlikely of spaces.

'Barletta'—diminutive onions

'Precovil'—the tiniest of peas

'Cherries Jubilee'—a yummy "new" potato

'Ronde De Nice'—Small, round zucchini (harvest at about 1")

'Peter Pan Green Scallops—1 to 2-inch scallop-edged squashes

'Sweet Dumpling'—mini squash

'Munchkin'—mini pumpkins

'Ruby Pearl'—thumbnail sized tomatoes

Alpine Strawberries—these are the fairy berries

Basil `Fino Verde Compatto'—a charming, compact mini-bush with tiny leaves.

A tussie-mussie of mini carrots from a child's mini garden

A Garden of Greens

When you hear the words vegetable garden, do you conjure up an image of a boring rectangle with straight rows of lettuce and carrots? Unfortunately, that is the image and the reality of most vegetable gardens. No wonder more children aren't thrilled by the prospect of planting vegetables. Let's try to do something just a little different, something that will captivate and personalize a garden for a child.

Find a flat, sunny spot for your child's garden. Turn the soil and add whatever is necessary to make it rich and loose. Trace out a large, plump heart with a stick. Dig a small furrow along the heart shape. The furrow should be about 1/4 to 1/2 inch deep. Inside the heart write your child's name in big letters. Again, dig another furrow of the same depth following your traced letters.

Choose hardy, fast-growing greens such as red and green leaf lettuce, cress, and, of course, the dependable radish, and plant them thickly along the furrows.

Let your child water the garden and watch the delight when the first greens pop through the soil. If you don't think that is exciting for children, you are

"Oh, come, come, come," HE SHRIEKED, *"I'm a father to a radish!"*

mistaken. My friend Georgie Van de Kamp described the response of a youngster in the Descanso Gardens Children's Garden. "Oh, come, come, come," he shrieked—"I'm a father to a radish!"

As your children harvest the vegetables, sow new seeds to ensure a continuous crop. Allow them to wash and prepare their own salads—they'll love them.

A Child's Own Rainbow

Rainbows are one of the most thrilling sights in nature. One moment a sky is dark with rain and in the next instant a gigantic arch with brilliant bands of color pierces the clouds and touches the earth.

Legends surround rainbows. Some Indian tribes believe that wherever a rainbow touches the ground, the brilliant colors give birth to the flowers of the field. We search for the pot of gold at the end of a rainbow, make a wish on one, watch the angels climbing the rainbow ladder into the sky parade, but most of all, like children everywhere, we stop, look, and feel a sense of wonder.

Help to keep the sense of wonder alive by creating a ground rainbow for your children!

Trace out the pattern of a rainbow arch and prepare the soil. Perhaps you could also trace a pot of gold at the end of your ground rainbow.

Take the children to the nursery and let them choose small plants in the colors of the rainbow: red, orange, yellow, green, blue, purple and violet. (Keep in mind the varying heights of plants and try to guide the children gently. After all, this is their rainbow). The pot of gold could be closely planted with

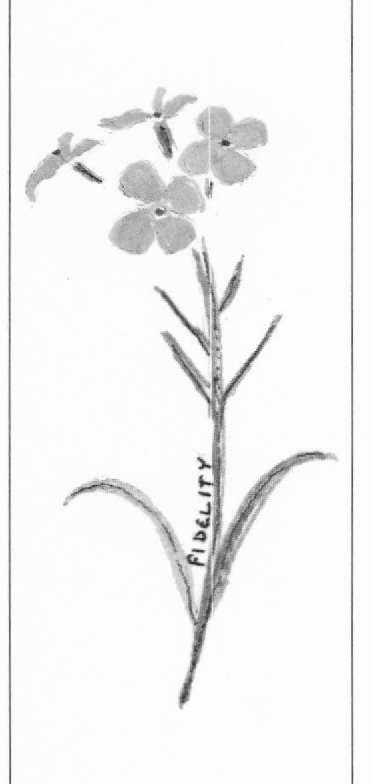

Make

me a rainbow

Make it soon!

I've been waiting all afternoon!

The raindrops heard

in their busy dance;

The sun shone out

and gave them a chance;

They seized the rays

with their fingers deft,

and wove the bright-hued

warp and weft.

SUSAN SWOOPE

one of the many varieties of tiny marigolds.

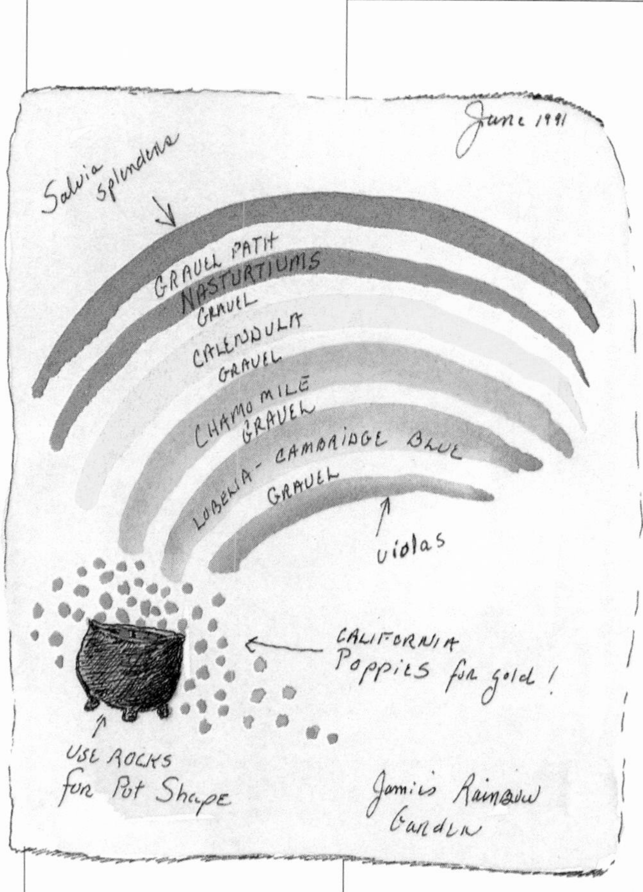

Spend a Saturday or Sunday afternoon working with your children in their rainbow garden. The secret word is 'with'. Show them how to gently tap the small plants out of their pots. Help them set their first couple of plants into the ground, firm up the soil around the roots and give the plants a light watering.

Take a "garden walk" every single day—rain or shine. Enjoy the enthusiasm that your children show; it's infectious.

When the rainbow garden is in its glory, take a color photograph of the children tending it. Pin the photo on a bulletin board or put it on your refrigerator and watch the kids smile whenever they look at their very own rainbow.

Longwood Gardens

What a joy to stumble upon the indoor children's garden that Catherine Eberbach and Mary Allinson designed for the conservatory at Longwood Gardens in Kennett Square, Pennsylvania. Crouching under the nasturtium bower and walking through the miniature labyrinth was like entering a different world. Everywhere I looked I found touches that would delight any child. Brilliant splashes of salvia, geranium, and nasturtium spilling over the walls and out of pots. A wayward hummingbird darting from bloom to blossom as a contented butterfly rested at the edge of a child-sized waterfall.

In the corner of a gazebo patterned after a spider's web, I surprised a little girl playing with a nasturtium doll she had dressed. Over her shoulder dangled a spider on a string. When I tugged it, a small bell rang somewhere above us. Around a bend a

young boy was sitting on a low, short bench decorated with paintings of Peter Rabbit's favorite carrots. In front of the bench, bricks were etched with the outlines of carrots.

In the place of a traditional wattle fence, the garden was walled with a pastel colored wattle with silhouettes of leaping rabbits leading the way through the maze of ivy hedges.

Huge, friendly topiary rabbits watched over the garden like sentinels as children crouched through hoops and passed through knee-high wooden gates and stopped to rest at a tiny picnic table and bench.

With my friends Louis and Virginia Saso and my husband Jeff, I wandered through the garden and lost myself in its beauty and humor. I could see Virginia posing in the arms of a giant topiary rabbit as Louis snapped her picture. Jeff was sticking his fingers in the bubble fountain (just like all the kids who entered the garden).

We all found it hard to leave the conservatory that day. The children's garden had captured our hearts and our imaginations and helped to awaken the sleeping child inside us.

EVERYWHERE I LOOKED I FOUND TOUCHES THAT WOULD DELIGHT ANY CHILD.

Precious Water

Every drop counts! Collect precious drips in a bucket or watering can. In my gardens I place terra cotta saucers under faucets—these collect water for my timid, ground-feeding birds.

Mulch! Covering the soil around plants with loose, protective material such as leaves, cottonseed hulls, compost, or straw conserves moisture and helps enrich the soil.

Water early in the morning so that the moisture can soak into the earth instead of evaporating into the hot, sunny air of afternoon.

Buy some rain barrels and catch the rain at each downspout.

Use soaker hoses or drip watering systems.

PLAN,

DREAM, WEED,

PLANT, HOE, WEED,

HOPE.

WATCH AS WATER

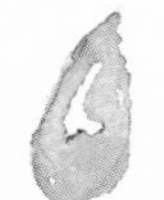

WORKS ITS MAGIC.

AND THE MIRACLE

UNFOLDS.

A Butterfly Garden

Noah and I quietly spread the blanket and stretched out on our backs. "Shhhhh," I whispered, "We have to lie still and not talk or the butterflies won't come."

A few minutes passed before the monarch butterflies began to land on the pineapple sage and sip its nectar. Soon, the sage and hyssop were studded with the brilliant orange of monarchs and one lone swallowtail. The butterflies sipped quietly until, disturbed by a quick movement, they rose (rustling like the thinnest parchment paper) into the cloudless sky.

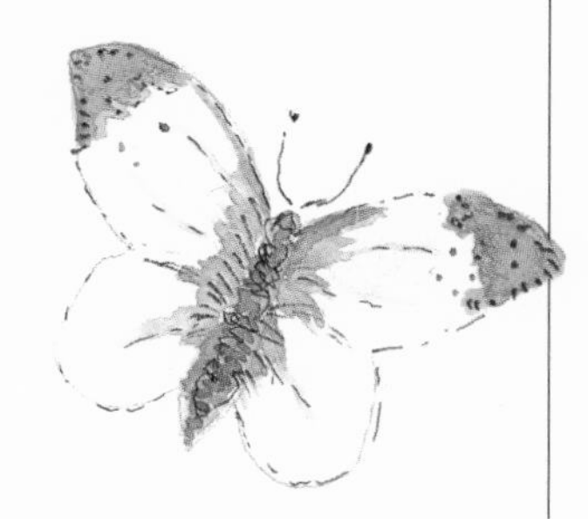

"Mom, the butterflies look like letters from the flowers, don't they?" Noah asked. I solemnly agreed.

As the afternoon winged past us, we spent time studying mosaic-bodied anise swallowtail caterpillars gnawing their way through a thick stand of fennel, and "woolly bears" (the mourning cloak caterpillar) feasting on a willow tree.

Late in the day, when the last of the butterflies had left their garden, we began the pleasant chore of watering the two small beds. Working in the dusky light, we noticed that our nicotiana and four-o'clocks, ignored during the day, were hosting half a

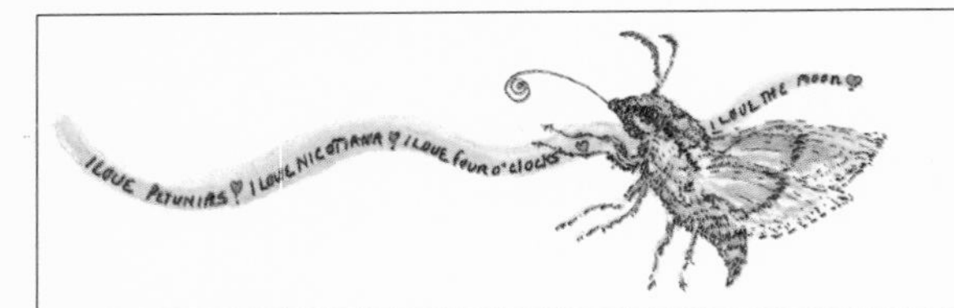

dozen hummingbirds. "No, Mom, look, those aren't hummingbirds, they're moths," Noah said.

We paused, still holding our watering cans, and waited patiently. The moonless night settled around us, the darkness studded with the glowing faces of mirabilis, nicotiana, silenes, night-scented stock and moon flowers. Silently, we watched and listened as

"MOM, LOOK, THOSE AREN'T HUMMING-BIRDS, THEY'RE MOTHS,"

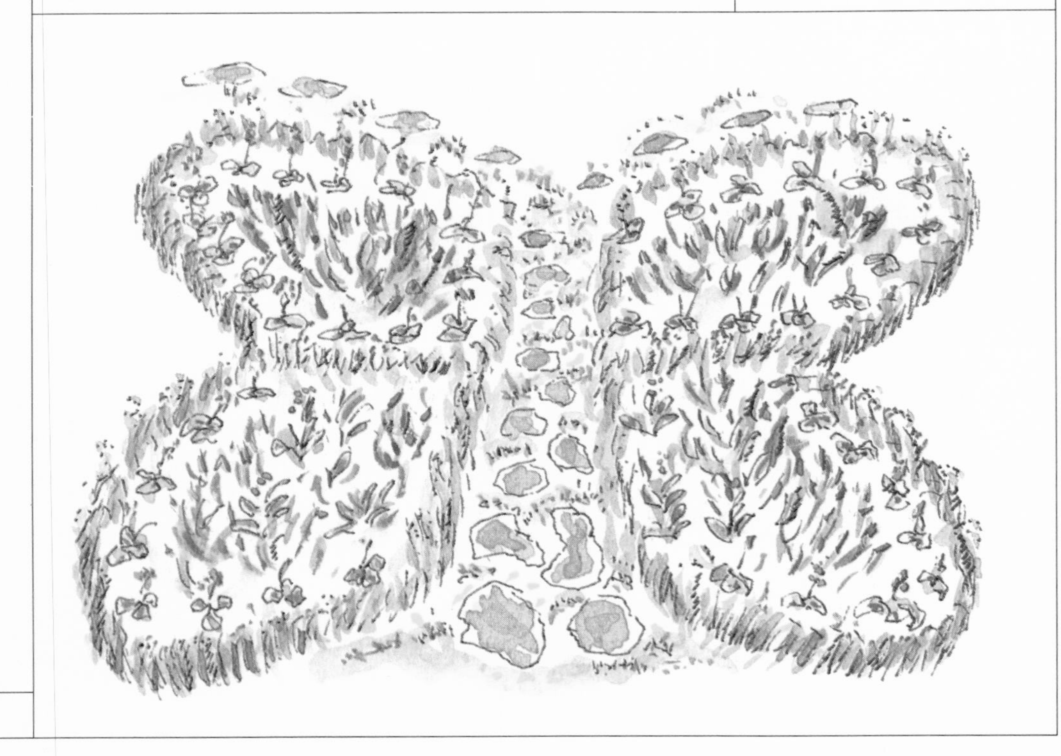

the cricket voices joined in chorus with the frogs and the faerie-like moths hovered (listen closely: you can hear their delicate wings) and fed at the blossom feast.

Noah looked up at me, his nutmeg-brown eyes dancing, "Mom," he said, "this is a miracle! It's like having our very own zoo and it's in our own front yard."

I guess in some ways, the small miracles we can create for our children are the ones most treasured and remembered. My small miracle of a butterfly and moth garden occupied about ten square feet and cost me only time, water and about $25 worth of seed, plant starts, and soil amendments.

The plan for our garden was simple. We chose a sunny site, protected from winds by a high blackberry hedge. We studied the shapes of butterflies wings and scratched the shape into our prepared piece of ground. Next, we outlined the butterfly with small rocks we had collected. We laid flat rocks between the butterfly wings; the rocks were the body.

Inside the butterfly wings we planted starts of perennial pineapple sage, lantana, jupiter's beard, hyssop, lavender, and coral bells. Between the stones of the butterfly body we underplanted with thyme and chamomile.

Brown and furry
Caterpillar in a hurry
Take your walk
To the shady leaf, or stalk,
Or what not,
Which may be the chosen spot.
No toad spy you,
Hovering bird of prey pass by you;
Spin and die,
To live again a butterfly.

CHRISTINA ROSSETTI

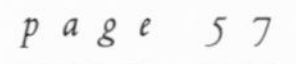

The rest was easy and fun. We went seed shopping, using a list compiled from years of my nature notes. Whenever I noticed a plant that was particularly favored by hummingbirds or butterflies, I would jot down its name and planting requirements.

We bought a White Swan 'Shake, Scatter and Grow' Butterfly Garden, too. This selection was a sure winner, developed by butterfly expert Dr. Robert Pyle, and extensively tested in White Swan's Oregon growing grounds. Our first test run with this seed selection was in a large terra cotta planter. The seeds germinated quickly, and as soon as the plants started blooming, myriad varieties of butterflies began visiting.

So now it is your turn. Buy your family a butterfly book and a magnifying glass. Spend some time outdoors in gardens or along country roads observing butterflies and the kinds of plants they visit for nectar and plants their caterpillars use for food. Plant a pot or plot for the butterflies and moths and wait patiently. You and your child will soon be rewarded by the quiet visits of "flower letters" of the darkness and the day.

PERHAPS *you would like to visit a special butterfly garden. Don't miss Butterfly World in Tradewinds Park, Coconut Creek, Florida.*

California hosts a Butterfly World in Marine World Africa USA, Vallejo.

Callaway Gardens in Pine Mountain, Georgia, has the Cecil B. Day Butterfly Center.

In Consideration of Faeries

When you are helping children plan their gardens, remember that the faeries and children must have a swath of low-growing greenery (it could be chamomile, thyme, or grass) on which to dance their dances and hold their faerie meetings.

During the day, the faeries are curled up sleeping in the hearts of the flowers. But at night, when the moon and stars are shining brightly, they troop out onto their green ballroom floor and dance and sing (with tiny cricket voices) in celebration of the children who gave them shelter.

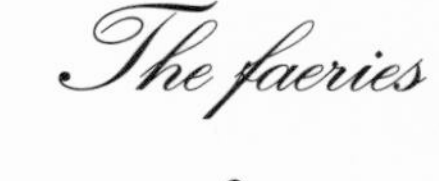

•

•

•

Janna of the Sunshine

Jane Hogue says that her 11-year-old daughter Janna is fascinated with the garden. "She explores barns and haymows, along the fence rows and ditches, down garden paths, and among the herb beds," she said. Janna reports home on any beautiful blossoms, unusual spider webs, flowers that are ready for cutting, ripe strawberries, and perfect hiding places.

Janna received a copy of Frances Hodgson Burnett's *Secret Garden* and immediately fell in love. She is busily making plans for her very own secret garden this summer, a place where she can weave dreams, adventures and sunshine into a beautiful tapestry.

Jane and Jack Hogue believe that raising children and growing gardens have many similarities. Both need nurturing and care to grow and blossom. Their children are thriving and blooming under their special care—we could all learn from them!

THE SECRET GARDEN

The Story of the Sunflower House

I love the sweet, sequestered

place,

The gracious roof of gold

and green,

Where arching branches

interlace

With glimpses of the sky

between.

ANONYMOUS

Working in my garden at Heart's Ease one day, I turned to greet an elderly lady. "Oh," she said, her voice full of nostalgia, "this reminds me of my childhood in Nebraska." I knew she must have some special memories to share. "Can you remember any special garden things you did as a child?" I asked.

She thought a moment then began a wonderful story: "We were poor and didn't have lots of store-bought things. My favorite flower project was our summer playhouse—we didn't have a regular play-house, but one we planted every year.

"In early summer, my mother would wake us up with 'Get up you sleepyheads, today's the day!' and we would get out of bed and pull on our clothes. We didn't even want to eat breakfast, but she would make us sit down and take our time. It all just served to heighten the excitement. We couldn't wait to get outside.

"Chores done, watering can and stick in tow, we would head outside and take time choosing the best, flattest, sunniest spot in our garden. Then the work would begin.

"Mother would use the stick to trace out a large rectangle, usually about 6 by 9 feet, leaving a small

My summer
home is the
fairest of all
With a morning
glory roof
and sunflower
walls!

LOVEJOY

opening for a doorway. She would drag the stick along the ground and gouge out a trench a couple of inches deep. My little sister and brother would trail behind and drop in seeds. John would drop in a big, fat sunflower seed; daintily, my sister would tuck in a 'Heavenly Blue' morning glory seed. I would trudge along behind them lugging the huge tin watering can. I'd use my foot to knock the earth back over the seeds and then I'd give them a small drink of water.

"Every day one of us would have the chore of walking that rectangle of land and giving a drink of water to the sleeping seeds. We all hoped to be the one to discover the first awakening green heads that poked through the soil.

"Once the green of the sunflowers peeked through the earth, we became even more interested in our growing playhouse. Usually, we would each water the plot once a day. Soon the sun-

GIANT SUNFLOWER CONTEST

"The competition for the largest head of sunflower seed, which closed on September 30, proved a greater success than we at one time anticipated. A few days before the competition closed, heads of sunflower seeds arrived from various parts of the British Isles...there were 140 competitors. First prize went to J. Golding, Summer Hill, Frensham, Surrey. The winner sent a magnificent black-seeded head measuring 52½" inches in circumference, probably of the variety 'Russian Giant'."

FROM THE GARDEN MAGAZINE *October 12, 1918*

Why not start a children's sunflower competition in your town? All entries should be harvested and measured by September 30th.

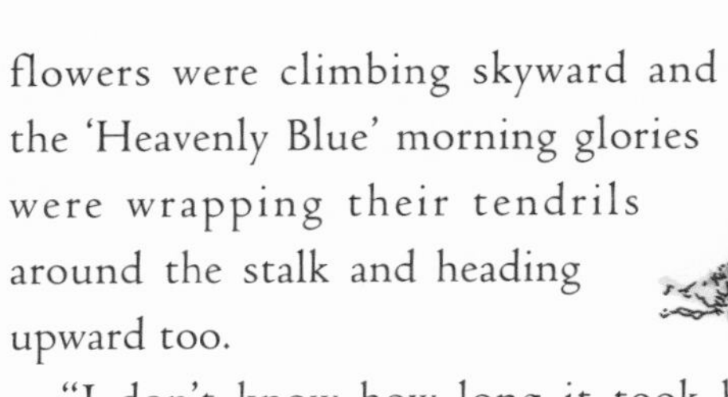

flowers were climbing skyward and the 'Heavenly Blue' morning glories were wrapping their tendrils around the stalk and heading upward too.

"I don't know how long it took before the sunflowers were at least twice as tall as us kids, but soon they were and Mother would come out with a big roll of used string we had saved up through the winter. 'John, you fetch the ladder and we'll get your roof going today,' Mother would say. My brother would drag out the big ladder and Mother would tie string to the top of one sunflower's neck. She would lace the string across that rectangle, back and forth, back and forth, 'til all we could see was a spider web of string against the blue Nebraska sky.

"In a matter of days, the Heavenly Blues would start journeying across the web, and soon the string was invisible. Looking up, all you could see was the gold of the sunflower faces, the green of all the leaves and like patches of the sky itself, the blue of those morning glories. I'll tell you there was nothing like crawling through the door of that playhouse and lying on the ground looking up through that incredi-

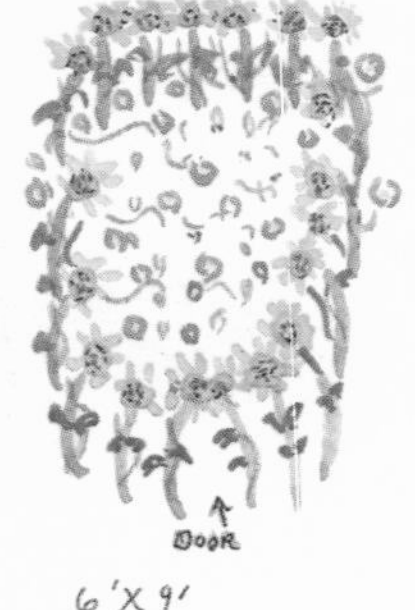

SUNFLOWER HOUSE TOP VIEW

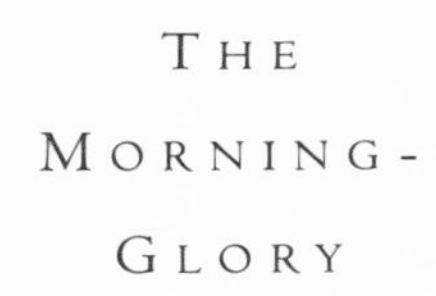

THE MORNING-GLORY

❖

Wondrous interlacement!
Holding fast to threads by
green and silky rings,
With the dawn it spreads
its white and purple wings;
Generous in its bloom, and
sheltering while it climbs,
Sturdy morning-glory.

from
De Gardenne Boke

ble lacework of vines and flowers. I guess you could say that I spent the best days of my childhood playing, dreaming, and sleeping in that little shelter."

❖ ❖ ❖

How do you go about furnishing a house so special? Surely you cannot go to a store and buy anything that will fit properly into such a home.

The children searched fields and woods and found everything they needed. Their mother did not have to teach them to do this; they just knew, instinctively, what was right for their playhouse. What they chose was what their mother and father, grandmother and grandfather had used before them.

Acorn cup & Saucer

Rosehip teapot
Thorn spout
Twig Handle

For a table, they rolled in a large, flat rock. Perfect chairs came from the woodpile—short, fat stumps. Doll beds were made of corn husks and down-stuffed milkweed pods. For carpets, moss and lichens; for coverlets, great big leaves (sycamores were soft, but woolly lamb's ears and old-fashioned mullein were the best). Dinnerware was not a problem—round honesty plant pods were dishes, acorns

and caps made cups and saucers, a plump, red rosehip poked with a thorn became a teapot with a spout.

Filarees were scissors, wild walnut halves were the porridge bowls (look at the heart inside them), beech leaves were napkins, and a burr-basket (from burdock) filled with miniature wildflowers sat in the middle of the rock-table. The garden was an endless toy store.

At night, the children ran barefoot through the grass catching fireflies. Gently, so as not to injure the fragile, flickering lights, they tucked them into the blossoms of hollyhocks and knit the edges together with a long twig. Some of the hollyhock-firefly lanterns were hung inside the sunflower house. Others were used in fairy-like processionals through the moist darkness of the garden.

"BOYS AND GIRLS,

COME OUT TO

PLAY,

THE MOON

IT SHINES

AS BRIGHT AS

DAY."

———

Lovejoy

Beans, Gourds, Pumpkins and Poles

Sweet summer dreams
In my runner bean tent
My friends who
played with me
Paid laughter for rent.

LOVEJOY

Children love a place of their own to use as a hide-out. I once saw this wonderful teepee—a perfect place for summer play. Here's how to make it:

Set four to six poles in the ground at an angle and bring them together at the top, securely lashing them with some heavy twine to form a teepee shape. Run twine roughly around the teepee to form a ladder for scarlet runner beans or showy painted lady beans (both flowers and pods are edible) and varieties of gourds. Plant seeds all around the base of the teepee. As the vines begin to reach upward, the children will be fascinated with the climbing process and the searching tendrils. (Show them how some vines always wind clockwise and others always wind counterclockwise).

As the teepee fills in, it becomes a secluded, dream and play-inspiring hide-out. An added treat is that the children can use the gourds they have grown to make bird houses, bowls, nests, toys—the possibilities are endless, and they grew them themselves!

Sand Castles

I SEE THE GARDEN THICKET'S SHADE

WHERE ALL THE SUMMER LONG WE PLAYED,

AND GARDENS SET AND HOUSES MADE

OUR EARLY WORK AND LATE."

MARY HOWITT

When I was a child growing up in the fragrant, golden hills of southern California, one of my favorite play areas was the wooden sandbox under our gnarled apricot tree. Flanking the sandbox were hollyhocks, carnations, iris, and rainbows of sweet peas—a palette of color for building castles.

Early dew-laced mornings would find me outside picking bouquets of blossoms to use in my construction project for the day. The night-moist sand was easy to mold and as I built, I studded the sides and roof with a mosaic of blossoms. Tips of bushes and ends of tree branches became a forest surrounding my creation. Tiny clumps of moss lined the path leading to the doorway—perfect bushes.

As the day progressed, friends joined in and helped add rooms and designs of flowers to the castle walls and pathways. At dusk, the last warning call came from our mothers—time to go inside now, or else! Deliberately and gleefully, we leapt into the center of our castle—every last vestige of our creation vanished. We felt no remorse, we knew our summer days stretched endlessly ahead. There would be other castles.

"That delightful thing, a sand-pit: it is a place of everlasting joy when one is small, and even when one is growing fairly biggish. We dig out arched recesses to sit in, and we build castles and all sorts of houses with the heap of loose sand at the bottom...And then we get some flowers and make quite a pretty garden round the house."

GERTRUDE JEKYLL,
from "Children and Gardens", CountryLife, 1908.

Write a secret message with lemon juice or milk. When it is dry the words are invisible. To read the message, press the paper with a warm iron. The words come back like

MAGIC!

Secret Notes

Ricky Beauclaire and I were the neighborhood whirlwinds. We were up in every tree, under every bush, and behind every rock. If there was trouble to be found, we found it (or maybe we caused it!).

We were inseparable, but when we were forced to be apart we found ways to communicate. Secret letters were tucked into a hole in our favorite old apricot tree and when Jackie Wingo found our hiding place we mystified him by writing with lemon juice "invisible ink". We put tiny, tiny messages inside the closed mouths of snapdragons, scratched letters onto mulberry leaves, and left cryptic trails of leaf messages on the sidewalk (which usually blew away before they were found).

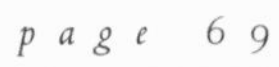

Faerie Tea Parties

"THE FAIRY QUEEN

WORE VELVET

CLOAKS OF PANSY

PURPLE IN SPRING,

WITH A PETTICOAT

OF LATE YELLOW

TRUMPET

DAFFODIL"

∴

The Garden Magazine

August 24, 1918

A recurring theme of faeries and woodlore cropped up in dozens of letters and reminiscences I received from throughout the country. And what would my own childhood have been without the secret hideaway of boughs in Grandmother's garden?

Dorothy Fitzcharles Weber, author of *Artistry in Avian Abodes*, wrote me, "Many of the nature crafts and lore I learned from scouting I practiced with my own children and then grandchildren at Crystal Lake in Maine.

"There were white birch trees, many unusual mosses, pink lady's slippers, curious rocks, hemlock cones, ferns and a multitude of natural materials.

"When Kim and Kelly were little girls we would have wonderful tea parties. The placemats were fern fronds, acorn cups were doll-sized tea cups, and then a choice of birchbark sandwiches filled with buttercup spread and tea brewed from soldier moss. Dessert was often pebbles á la mud."

From many miles away came this note from Susan

To the fairy land afar
Where the Little People are;
Where the clover tops are trees,
And the rain pools are the seas,
And the leaves like little ships
Sail about on tiny trips.

H.T. JOHNSON
From The Garden Magazine
May 18, 1918

Deep in the wood
I made a house
Where no one knew the way;
I carpeted the floor with moss,
And there I loved to play.

I heard the bubbling of a brook;
At times an acorn fell,
And far away a robin sang
Deep in a lonely dell.

I set a rock with acorn cups;
So quietly I played
A rabbit hopped across the moss,
And did not seem afraid.

That night before I went to bed
I at my window stood,
And thought how dark my home
must be
Down in the lonesome wood.

K. Pyle

Jones Sprengnether, who grew up in St. Louis, Missouri.

"I clearly remember making homey places for the faeries who lived in the woods. These places were snuggled in amongst the roots of trees and mostly in mossy areas. We made dishes of acorn tops, cradles of walnut shells, leaves were cots or hammocks, and tables were made of twigs lashed together."

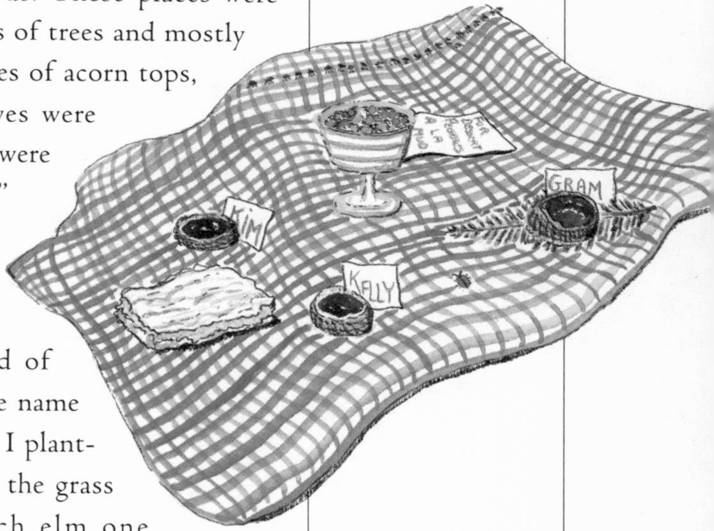

And from many, many years away:

"It was to please the faeries that, long before I had heard of naturalizing bulbs or knew the name and the fame of W. Robinson, I planted a ring of white crocuses in the grass round the bole of the wych elm one November. When the white circle appeared in spring, Mother said, 'Who put those crocuses in the grass? It must have been you, Nan.' I hung my head, expecting a wigging, but Mother smiled and said, 'Your grandmother used to do that in Bethnal Greens in 1820. What made you do it?' But I did not tell her, nor anyone else, that it was to please the faeries."

Poppy Maidens

Gretel Hanna Barnitz sat in my Heart's Ease gardens and shared childhood memories from the 1920s. As a child, she loved to sit beside her mother's shirley poppy bed and carefully turn down the silky petals of each poppy, exposing the little green seed pods for heads. Then, she would tie a blade of grass around their "waists", creating graceful, dancing, poppy maidens.

"Some afternoons we would spend hours carefully turning and tying all of the poppies in Mother's side garden," she said.

"One day Mother called me inside and as I turned to take a last look at my poppy ladies a breeze started stirring through them. All of the poppies began bowing and waving as though they were saying, "Goodbye, thanks for a great day of play!"

Poppy maidens
Bowing, swaying
Watched the girls
As they were playing.

Lovejoy

LOVELY DANCING GIRL

Please play with me

I'll make you leaf boats

To take you to sea

MY FILAREE SCISSORS

will cut you a shawl

Of dew studded spider webs

Dressed for the ball!

A HAT FROM A TINY PETAL

of blue

A peek of your anther

We'll call it your shoe!

A HOLLYHOCK DANCING SKIRT

Hearts ease for a face

Twigs for your arms—

Now, your hair's a disgrace!

LOVEJOY

Daisy Grandmothers

Years ago, I quizzed an elderly lady who still spends most of her time in her garden. My usual question, "What did you do with plants when you were a child?" was met with a bright, quick answer.

"We made daisy grandmothers," she said, smiling as she nodded her head in thought.

"Daisy grandmothers?" I asked, "I don't recall hearing of those before."

"Oh," she said, "that wakens fond memories of spring, when we four sisters would take scissors and a pencil and head out to the meadows searching for the best, fullest field of daisies."

She explained how the girls would spread out in a meadow and pick a handful of daisies and then "trim their bonnets". "We would work around the face of the daisy, cutting the petals in the shape of a bonnet," she said. "Generally, we left two long petals at the bottom of the daisy face. Those would be the bonnet ties. Oh, what great fun we had with our daisy grandmothers!"

Years later, my friend Christine Nybak Hill recounted tales of making daisy grandmothers out in the fields, too. Somehow, through time and conti-

SOMEHOW, THROUGH TIME AND CONTINENTS AND GENERALLY WITHOUT BENEFIT OF BOOKS, OUR FLOWER TRADITIONS ARE PASSED ON FROM GENERATION TO GENERATION.

nents and generally without benefit of books, our flower traditions are passed on from generation to generation.

I found this poem in a children's garden book from the late 1800s:

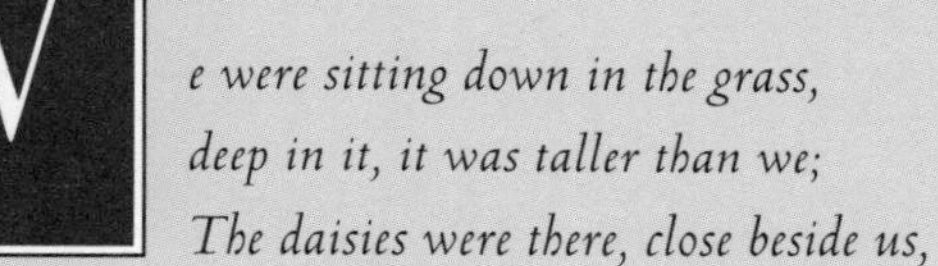

We were sitting down in the grass,
deep in it, it was taller than we;
The daisies were there, close beside us,
In a circle they stood on a mound,
And Auntie took out her sharp scissors
And she snipped them around and around,
Until each had a white cap border,
And she left them two petals for strings;
And then next she found a lead pencil
In her bag with the rest of her things;
And with that, on each yellow center,
Auntie drew such a queer little face—
But look! You can see the grandmammas,
Here they are in the same grassy place!

Hollyhock Dolls

Hollyhock dolls were the plaything most commonly mentioned when I quizzed people about childhood play. "Oh, yes, my sisters and I always made hollyhock dolls," one lady told me. "There was an unlimited supply of hollyhocks next to our barn. Every morning we would go outside and pick buds and blossoms. We used them stacked on top of each other and held together with twigs. We loved to make them cloaks of violet leaves sewn together with thin strands of grass.

"Our big brother

HOLLYHOCKS

The hollyhocks greet you

Wherever they meet you,

With stiffest of bows

KATHERINE H. PERRY

made dolls for us using small cherry tomatoes as heads, but we liked our dolls to be all flowers. Canterbury bell blouses, petunia skirts, these were some of the other flowers we used."

"IT IS FUN TO GO OUT VERY EARLY in the morning when the air is still cold ", wrote Stephen Law of Friendship, Maine. "Find a Hollyhock with a big, fat, black and yellow bumblebee sleeping in it and pet it!" (I've always caught them in the cup of my hands and have never been stung—the secret seems to be to handle them gently and let them just crawl about.)

Trumpet Flower Dolls

"THE MORE THINGS THOU LEARNEST TO KNOW AND ENJOY,

➢➢➢➢➢➢➢

THE MORE COMPLETE AND FULL WILL BE FOR THEE THE DELIGHT OF LIVING."

PHALEN

Grab yourself some trumpet flowers and leaves and we'll work on a doll for you," Gram said. I reached out and picked two blossoms and a handful of leaves.

Gram turned the bloom with the narrow end up, and stuck a sunny dandelion face into the mouth of the trumpet. Next, she bent a fresh, green leaf over the head and pinned it with a twig. "There, now she has a happy face and a beautiful bonnet," she said. Finally, we fashioned her a cloak out of one of the leaves, and a peach tree furnished her with twig arms. Gram made a dainty necklace of tiny pink clovers.

"Let's make her a sister," I said. "Let's dress her like a Hawaiian hula dancer," Gram chuckled, as we tied a long strand of grass around the dancer's waist. Then, we draped many strands of grass over the waistband and we had a hula skirt.

"Let's give this gal some dandelion curls on a rosehip head," she said as we split some dandelion stems, sucked on them, and made curls.

Our trumpet vine dolls danced and played all through a rainy, summer day. In the evening we took them down to the swollen creek and let them float away for an adventure of their own.

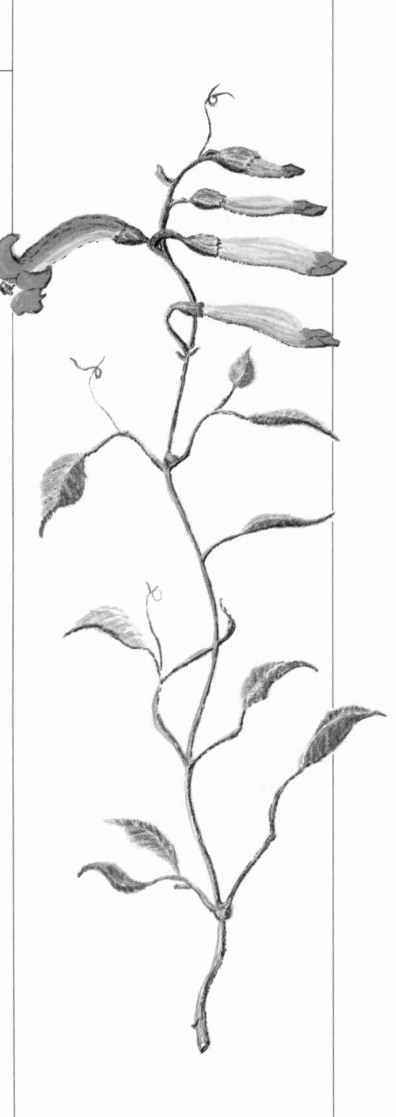

Pansy and Heart's Ease Dolls

In the 1800s, young ladies would often sit for hours and cut out strings of headless paper dolls. The paper was of two thicknesses; in other words, one piece was folded in half lengthwise, and the dolls cut out with the fold at the top. Then our Victorian friends would slip pansies or Heart's ease blossoms into place for heads, and the little ladies would be able to stand on their own. When paper was not available, leaves would suffice. They would simply pick a pansy or heart's ease, fold a leaf around the stem like a cloak, and poke twig arms through for a little pansy leaf doll. A leaf also could be folded in half, like a single paper doll, a hole slit for the head, and a pansy on a stem stuck into the hole.

PANSY FACES

Each pansy has a smiling face
To greet me when I go
To work among them with my spade,
And help to make them grow.

LOUISE MARSHALL HAYNES
Over the Rainbow Bridge

Trees for Wishing, Trees for Dreaming

More than forty years ago a little boy with rosy cheeks and curly, cowlicked hair found his own tree. He didn't know that it was a sycamore tree, over a hundred years old, the last of dozens which had once arched over Wildcat Creek.

The tree became the boy's friend. When childhood problems overwhelmed him he ran to the tree, scrambled up its welcoming branches, and found comfort in its huge, sheltering presence. After a few years the one-sided conversations he had with his friend became part of a daily routine. The dependable old sycamore was always there to hear his wishes, dreams, and sorrows.

On the day his family left Indiana to move to California the little boy made one last trip to his friend. He hugged his tree and as he walked away, a strong wind stirred a piece of paper he had left in a hiding place. In childish writing the paper said, "I, John Arnold, love this tree."

John's Buttonwood tree

My Tree

Spreading wide
her skirt of leaves,

Stroking the wind,
Tickling clouds,

Chatting with
birds,

Sheltering children
on friendly boughs.

Lovejoy

A Tree For My Son

On the day I found out that I was pregnant, I bought a tree for my unborn son. During the months that Noah was growing inside me, the tree grew and flourished.

Afraid to plant the tree in the ground and leave it behind in a move, I planted it in a large terra cotta pot. To this day the Deodar cedar grows in a pot, gets fed regularly, and is talked to and nurtured like a child.

Give your child a birth tree. As the child grows, the tending of the tree can be a favored chore. The growth of the tree can be charted easily by tying a bit of colored string to the outermost tip of a branch each fall. During the spring and summer growth can be measured weekly.

DEAR LITTLE TREE that we plant today
What will you be when we're old and gray?

THE TREE ANSWERS

"The savings bank of the squirrel and mouse.
For robin and wren an apartment house,
The dressing-room of the butterfly's ball,
The locust's and katydid's concert hall,
The schoolboy's ladder in pleasant June,
The schoolgirl's tent in the July noon,
And my leaves shall whisper them merrily
A tale of the children who planted me."

UNKNOWN

What Dreams Are Made Of

THE HOUSE OF THE TREES

"Lift your leafy roof for me, Part your yielding walls: Let me wander lingeringly Through your scented halls."

Ethelwyn Wetherald

Sunday mornings are my favorite time. I can curl up in bed and read and listen to the bird voices outside my window. I love to hear the quail coveys pattering across the roof—they sound like gentle rain.

Sunday mornings are when I do a lot of research from old garden and nature books. These are times of quiet discovery. Whenever I find a great garden poem, a story about children in the gardens and fields, or a story of a special garden, I feel as though I have uncovered a treasure.

Today I stumbled across an old etching of a Victorian garden. The etching depicted a huge catalpa tree. The tree had been pruned into the shape of a house—a true tree house! Windows and doors were simply cut through the outer leafy branches of the tree and the furniture was boughs. I could imagine half a dozen children eagerly climbing a rope ladder and entering the leafy bough house for an afternoon of sun-dappled dreaming and play.

The Need For A Swing

My friend Georgie is in her early 80s and when I am with her, really, I think of her as a very, very young person. We were out in her garden last week admiring swaths of blooming, golden daffodils under her huge, centenarian oak tree.

As we stood eyeing the daffodils she looked up into her tree and said, "I really think every child should have a tree swing, don't you?" I had to agree; I think a tree swing is one of the simplest and most wonderful garden pleasures.

"I used to love the freedom of swinging, and now that I think of it I really think that I NEED a swing in that tree. Yes, I NEED a swing," Georgie said emphatically.

I have always heard that there is a bread-and-water type of necessity and then there is soul-food necessity. I think that Georgie hit upon a definite soul-food garden necessity—a swing. It could be a tire hanging from a tree, even a plain rope hanging Tarzan style, or a swing with a wood plank seat.

However you do it, remember what it felt like to have the wind rushing past your ears as you tried, pumping harder and harder, to reach the very spot where the green earth ended and the blue sky began.

How do you like to go

up in a swing,

Up in the air so blue?

Oh, I do think it is the

pleasantest thing

Ever a child can do!

ROBERT LOUIS STEVENSON

A Living Gazebo Playhouse

Curtain of

SHIMMERING

leaves

I was wandering through the fragrant, meandering pathways at Lewis Mountain Herbs and Everlastings in Manchester, Ohio. As usual, my head was down, my eyes darting about at the variety of herbs and scented geraniums.

As I came to a Y in the path I looked up, trying to decide which fork to take, and gave a start of surprise. Ten feet in front of me was a lush, living gazebo of Ohio Melrose apple trees. The bough-canopied gazebo was punctuated by the straight trunks of the trees and festooned with slender arching branches. I headed for its sheltering circle.

I spent most of my afternoon sitting inside the gazebo surrounded by a curtain of shimmering leaves. A companionable brown towhee (we call them brown bettys in my family; the name seems to suit them) scratched through the soil, apple tree shadows playing across the ground and over her back.

To this day I think back to my afternoon in Ohio and realize what a haven that apple gazebo was for me. Imagine what it might mean to a child to find such a whimsical and playful space in the middle of an otherwise hands-off, grown-up garden.

The construction of the apple-tree gazebo was simple. John Lewis cultivated a circle and planted eight trees around its perimeter. Each one of the trees was sturdily staked to keep it straight. As the trees branched out, John pruned them to the shape he desired. Long branches were tied or wired to each other to form the arching canopy. Moderate trimming throughout the summer keeps the gazebo neat and inviting. I can't wait to plant this playhouse!

S
N

Crowns

A peculiar connecting thread runs through all the garden stories I've collected. Somehow, though fifty or sixty years might have passed, the child inside each person lightens and glows through ageless eyes as they tell their stories.

One sunny afternoon in May, I was sitting on the stone patio writing. A familiar voice greeted me and as I squinted into the sun I saw a snowy-white halo of hair. My friend, Millicent Truax Heath, sat down on a stool next to me. Millie's eyes brightened and her voice quality changed as she began telling me about her childhood summers in Rockport, Minnesota.

"My Aunt was Sarah E. Truax. She was a charter member of the San Diego Art Guild and she was an artist in everything she did. She was a wonderful lady! I remember what fun we had with her in the fields. She would take her easel out to paint and we children would go with her. She really made us look at nature and we would lie on the ground, peering through tangles of wildflowers trying to find the little people we knew were there. Sometimes she just put a crumb down on the ground and we watched, fascinated, as the ants picked up that boulder of a

Ivy and Roses
Woven around
Mary's a princess and
This is her crown

Clover nasturtiums a
Bright passion flower
Sarah's a gypsy queen
This is her hour

A Tiara of Fireflies and Flowers

"Our Grandma always wore hair nets to keep her silver hair in place," a friend told me. "As the nets wore out or snagged, Gram would give them to us girls.

"On special evenings my sis and I would weave garlands of Daisies and wildflowers and pin them in our hair. Our brother would bring in a canning jar full of fireflies and we would tip it into Gram's hair nets, loosely fit them over our fancy hairdos, and have twinkling tiaras for an evening of garden play."

crumb and carried it away. We all got so much enjoyment out of everything!

"Aunt Sarah would pick a beautiful bouquet of the long stemmed pink clover that grew in the fields. Then she would make us crowns, necklaces, bracelets, and rings from clover braids."

Millie leaned over and picked some clover out of a patch of weeds. She laid three across my lap and began to braid them. As she braided, she added more blossoms until the braid was long enough to encircle a head. Then she poked the loose stems into the beginning, took a long stem of grass, and neatly tied the ends together to make our crown.

Millie said that once she and the other children learned how to make the braids, everything was fair game. They would stop alongside the road and braid bracelets and necklaces and stud them with violets, hepaticas and different colors of clover. They wore them everywhere—even to church.

Millie's sweet, clover-filled memories of seventy years past came alive for me as I learned to braid my own crown. Just as I was tying the ends together a young towheaded girl asked what I was doing. Before I knew what was happening, she was sitting on the stones, surrounded by a crazy quilt of flowers, slowly,

laboriously, braiding her own crown. "Look, it's perfect," she said holding it up for all to see. "That's not a real crown," her older sister scoffed. "Yes, it is," she said, "it's the important kind!"

More Crowns

A few years ago Robert Ball, a former botany professor, told me the story of yerba buena crowns.

"When I was a child, our family would go on overnight camping trips," he told me. "My father always had us gather long strands of yerba buena and twine and weave them into crowns. We wore those to bed not only for fun and the good smell, but to ward off pesky mosquitoes."

Growing up in Grandmother's garden, the changing seasons brought me an endless variety of playthings. In the spring, I would string tiny Cecile Bruner rose buds to make necklaces and garlands. In the fall, I would string wild rose hips. I never realized that anyone else had ever thought to do this until I read what Gerard wrote in the late 1500s. "Children with delight make chains and pretty gewgaws of the fruit of roses."

YERBA BUENA
PICKED TODAY
HELPS TO CHASE
THE BUGS AWAY.

LOVEJOY

Clover Chains

SIMPLE

pleasures

LIFE'S

treasures.

Several years ago, I planned a day in the garden with two young friends, Sarah and MaryBeth Monger. As I walked up the pathway to their Grandma's house, I noticed Sarah sitting in the grass, head bent over her lap as she worked nonstop on something.

She was so engrossed in what she was doing that she didn't even see me watching her. Picking a clover, she would slit the stem just below the flowerhead with her thumbnail. Then she would push another clover stem through the slit until the head stopped it from going further. Doing this procedure over and over, she fashioned a five-foot-long rope of clover blooms.

Sarah looked up with a glowing face and said, "Look, Sharon, a day-long jump rope." And that is exactly what it was, lasting Sarah and Marybeth all afternoon. It was a simple pleasure, made fresh for each day of play.

Daisy chains, clover chains, flower braids, and wreaths—the simple, repetitive steps in making these timeless delights are known by children everywhere!

Sarah's Jumprope

Sarah waited patiently

Stringing clovers 1-2-3

Clover jump rope

For her play

It will last for

Just one day!

Lovejoy

Leaf Hats

In poking through old books of nature crafts and pastimes for children, I found a story of hats of leaves pinned with twigs. I planned to tell MaryBeth and Sarah about leaf hats, but they surprised me as I walked through the gate to their grandma's garden another time—they were both wearing hats of leaves that were pinned together with small twigs!

Sarah's hat was a large, elaborate leaf decorated with sprays of spiraea and tied under her chin by the stems. Mary's hat was smaller. She had used the stems to tie the leaves together in an overlapping, flower petal effect. Her hat was gaily decorated with sprigs of flowers inserted through slits in the leaves. (Imagine how beautiful and festive an autumn leaf hat would be!) The girls wore their hats most of the day, and when they were through with them they turned them into boats for their fuchsia dancing dolls!

During the long, summer afternoon, Sarah and MaryBeth foraged through Grandma's garden for more dress-up supplies. MaryBeth picked snapdragons from a tall stalk and used the individual "snaps" as clip-on earrings. Sarah chained clovers together to form a crown and as she worked, she tucked in nasturtiums

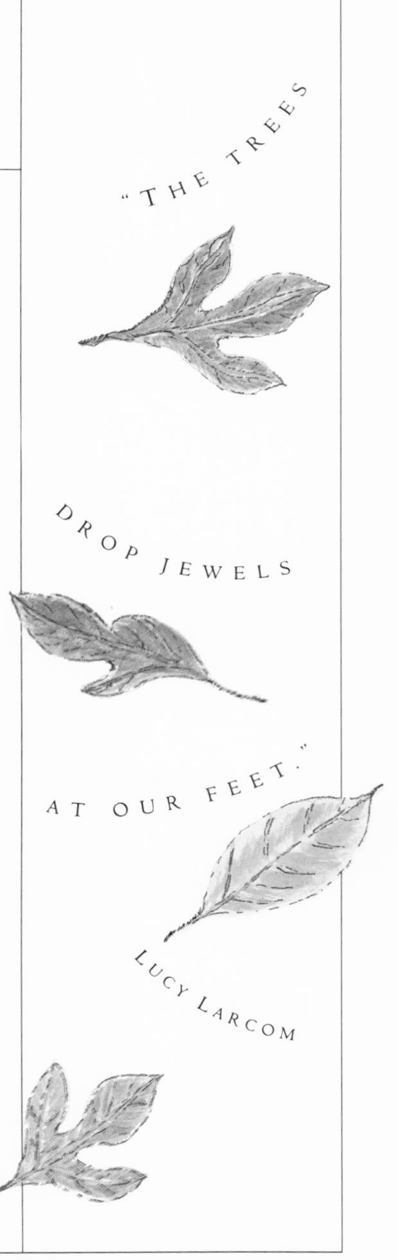

"The trees drop jewels at our feet."
Lucy Larcom

and passion flowers for "jewels." She looked like a flamboyant gypsy queen when she finished.

MaryBeth twined ivy and honeysuckle into a circlet and as she twined she wove rosebuds, forget-me-nots, and clover into its rim, she became a fairy princess crowned with rubies, sapphires, and diamonds.

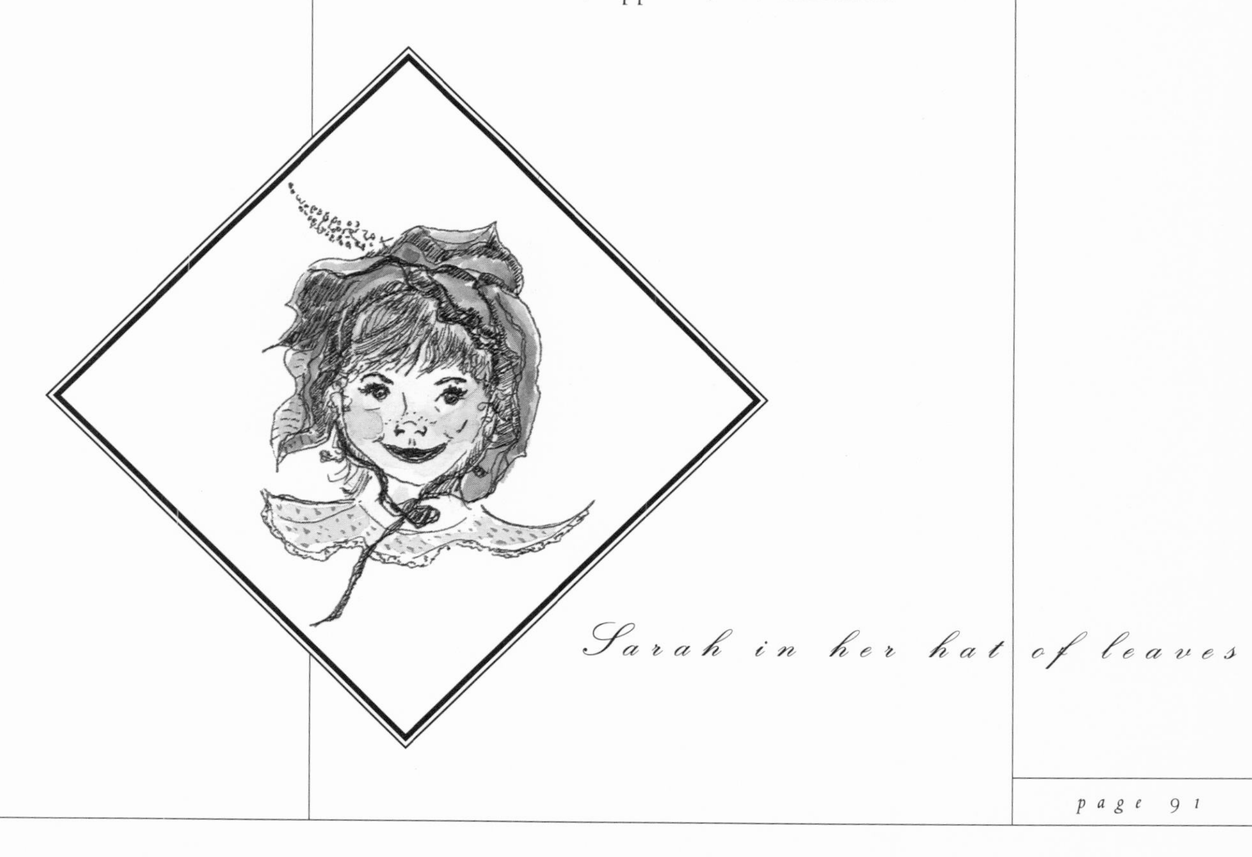

Sarah in her hat of leaves

Sassafras Chains

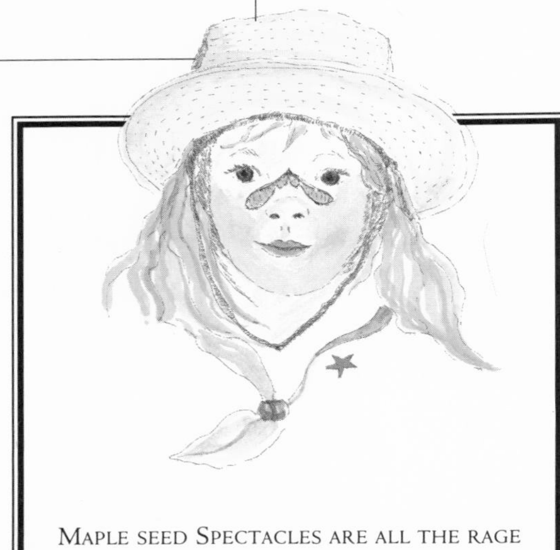

Maple seed Spectacles are all the rage with the fashionable set.

"Although the years of the Second World War were tumultuous, somehow I remember them as the peaceful foundation of my life," Patricia Reppert told me. "My father was a surgeon who joined the service and was gone for five long years. To provide me and my brother with a home, my mother packed us up and we moved to my grandparents' farm, Whittenoak, in Charlottesville, Virginia.

"I remember that my father's letters were anticipated and treasured by all of us—he wrote to mother almost every day. It seemed as though our lives revolved around the daily mail delivery. "Every day, during nice weather, my mother, brother, and I would hike through the fields and wait alongside the road for our mailman. You see, the mail was our life-line.

"The roadsides were beautiful, a miniature, flower-filled garden, perfect for kids. The woods skirting the road were filled with bushes, and our favorite, the wonderful sassafras tree! Sometimes we would spend several hours waiting, and during that time we made flower and leaf toys to keep ourselves happy.

"My favorite things to play with were the sassafras leaves. They were large, almost hand shaped, and fun to make into chains and clothes. We would start by pinching the long stem off the first leaf. From then on we would pick another leaf (leaving this stem on) and we would overlap the leaves, using the long stem like a needle, piercing the leaf in small stem-stitches. They really did seem sewn together, and we would make vests, over-the-shoulder Carmen Miranda-type blouses, crowns, belts and long, long chains. I remember one chain that was about 15 feet long. My brother and I tried to walk all the way home without breaking the chain. We never tired of playing our leaf games!

L • E • A • V • E • S

The orange-tinted sassafras

With quaintest foliage

strews the grass;

Witch-hazel shakes her gold curls out,

'Mid the red maple's flying rout.

Mayday

May is such a joyous month for me. My gardens are brimming with new life. The perennial borders are a riot of color, towering hollyhocks, gray carpets of carnations, cerulean delphiniums, candelabras of foxglove, and everywhere columbine, the pastels of the little dove flower amidst the lacy foliage of love-in-a-mist. And my roses! Many of my antique roses are at their best. The scents from every corner of the garden are overwhelming.

I feel like a child again—that same quickening inside that I felt every May Day when I filled tiny baskets and cornucopias with flowers, hung them on my neighbors' door knobs, rang the bell, and ran for my life. Crouched down amongst huge bushes of sweet smelling, old-fashioned geraniums I would peek out, heart pounding wildly, and watch as Mamam Braden, Goldie Pickering, and old Mrs. Downs found their baskets. (Perhaps some of the flowers in the baskets were from their own gardens!)

May Day

If I were asked the season,
I could not tell today;
I should say it still was Winter—
The Calendar says May.

If this indeed be May-day,
I must be growing old;
For nothing I was used to
Do I today behold.

On May day in New England,
In that old town of ours,
We rose before the daybreak,
And went and gathered flowers.

And then in pretty baskets,
With little sprigs of green
We placed them, and stole homeward,
And hoped we were not seen.

RICHARD HENRY STODDARD

A simple gift of flowers filled me with joy—and I know now that my gift gave joy to some lonely neighbors. And today, spending time with people who tell me stories of their childhood, May Day always looms brightly in their memories. Their eyes light up while remembering May garlands, May poles, and secret May baskets left for friends and loved ones. And the question most often asked: "Whatever happened to May Day?"

I plan to find May Day again. This year, and every year hereafter, I will leave a May basket for someone who may be lonely and isolated. I want to feel that excited joy that made my heart pound as I crept up to doors and left my small gift of flowers.

MAY DAY MORNING

Oh, let's leave a basket of flowers today
For the little old lady who lives down our way!
We'll heap it with violets white and blue,
With Jack-in-the-pulpit and
Wildflowers too.

We'll make it of paper and line it with ferns
Then hide – and we'll watch her surprise when she turns
And opens her door and looks out to see
Who in the world it could possibly be.

VIRGINIA SCOTT MIKE

Wishes and Charms and Four-Leaf Clovers

An even-leaved

ash,

And a four-

leaved clover,

You'll see

your love,

'Fore the day

is over.

Even ash, I thee do pluck,
Hoping thus to meet good luck.
If no luck I get from thee,
I'll wish I'd left you on the tree.

White rose, white rose,
Bring me good luck.
Good luck to you, good luck to me,
Good luck to everyone I see.

Daisy Divination

One for sorrow,
Two for joy,
Three for a girl,
Four for a boy,
Five for diamonds,
Six for gold,
Seven for a secret
Never to be told.

Find a four-leafed clover and pick it up,
All the day you'll have good luck.

Down among the meadow grass,
Searching it all over,
What a merry band are we,
Hunting four-leaf clover.

I wish, I wish is what you say as the dandelion blows away!

ONE I love, TWO I love,
THREE I love, I say,
FOUR I love with all my heart,
FIVE I cast away.
SIX he loves, SEVEN she loves,
EIGHT both love,
NINE he comes, TEN he tarries,
ELEVEN he courts, TWELVE he marries.

I know a place where the sun is like gold, And the cherry blooms burst with snow,

And down underneath is the loveliest nook Where the four leaf clovers grow.

A clover, a clover of two,
Put in your right shoe,
The first young man you meet,
In field or lane or street,
You'll have him or one of his name.

THREE WHITE ROCKS,
AND THREE RED BERRIES,
THREE YELLOW DAISIES,
FROM THE FIELD.
OVER YOUR SHOULDER
INTO THE STREAM
YOUR LOVE WILL VISIT
WHEN NEXT YOU DREAM.

Periwinkles

Sometimes we • tend to overlook • what's right • under our noses • Such was the case • with periwinkles

The soft, azure, star-like blooms of the periwinkle give the only color to be found in the shady recesses of the hillside below my garden. Even on the darkest California winter day I can usually depend on this faithful bloomer to enliven my herbal bouquets with a dash of quiet blue.

Over the years, I've asked flower lovers all over the United States and Britain for legends and lore about the humble periwinkle, but always to no avail. Sometimes we tend to overlook what's right under our noses. Such was the case with periwinkles and my friend Margaret Harper.

Margaret arrived one day smiling impishly, carrying a small bouquet of periwinkles. "Do you know the story of Perry Winkle's Paintbrush?" she asked. "It was told to me by Rossie Fairbairn. Her mother, Isabelle Evans, of Blue Lake, California, kept Rossie entertained when she was a child with stories of the wonders of plant life. Now, I share the story with my grandchildren, who never tire of it or of the search for the paintbrush."

Margaret set her bouquet down on a table and pulled out a single stem in bloom. As she began

D3ar fairi3s,
Grandma told
m3 that you
only r3ad l3tt3rs
writt3n with
p3rry's paintbrush.
So, I am using
on3 now!
I saw you
dancing with
th3 fir3fli3s
Last NiGHT!
Lov3,
N

telling the story, she carefully and slowly removed petal after petal.

Perry Winkle's Paint Brush

The first Spring descended upon the earth, and all of the new, young animals and shimmering green plants and trees were healthy and happy. The view from a hilltop across mile after mile of wildflowers with all different blossom and leaf shapes revealed one startling, glaringly obvious thing. Somehow, the finishing touch had been overlooked! All the flowers were one color—WHITE!

The very last flower to have been created was the humble periwinkle. Thus, he was the one called upon to solve the problem of coloring all of the flowers.

"Goodness, gracious," said Perry in a small, blue voice. "I am depressed. There is just no way a little flower like me could color all the flowers in the world."

"Perry," a deep, soft voice resounded, "what you need is a little faith! In this world everyone has a job and is expected to work. The job for you and your family will be to paint all the flowers every color to be found on our earth and in our sky."

In the language of flowers, the periwinkle is the symbol of pleasures, of memory and early friendship.

"But how can I do such a thing? There are not enough brushes or paint in the world to color the millions of flowers you have strewn on this planet," Perry said in a defeated tone.

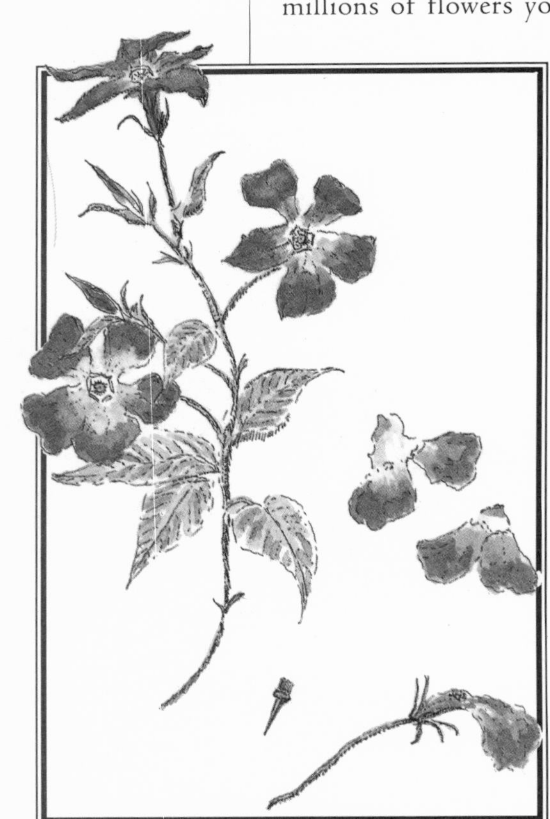

"The rainbow will be your never-ending supply of colors. And listen closely: Slip your petals off, and you will see that I have given every periwinkle in my kingdom its very own paintbrush."

As Margaret ended her story, she slipped off the last pale blue petal and pulled out the tiniest, most fairy-like paintbrush I have ever seen. Since that day with Margaret, I have introduced dozens of children to the magical, hidden secret of the often-overlooked periwinkle.

Pansies

"When God

created

humankind, he

gave each person

two eyes, but

only one mouth.

Two eyes to

appreciate

doubly the

beauty around

us."

Nikolai Reimer was a nurseryman and plant lover who grew up in Czarist Russia and emigrated to Canada. His daughters Mary and Frieda say, "In the beginning, our father neither understood nor spoke the English language, but he was an expert in the universal language of flowers." Nikolai founded the now well-known Reimer's Nursery in Yarrow, British Columbia. When people in the area suggested that he spend more time growing fruit trees and other useful plants—plants that would provide food to eat—he would say, "When God created humankind, he gave each person two eyes, but only one mouth. Two eyes to appreciate doubly the beauty around us." He propagated great numbers of trees and shrubs, but flowers remained his first love.

PANSIES

"I am thinking of you," is what the pansies say
When they come to you from a friend;
And, "I am thinking of you" is what they say
When you the blossoms send.
No need of words when pansies are near
To carry the message for you—
Just send a bunch of the blossoms fair,
They'll speak plainly as you could do.
All over the world in their simple way,
No matter where they go,
"I am thinking of you" is what they say,
And all people their language know.

Margaret Coulson Walker

"Our father taught us many things about plants and flowers," say Frieda and Mary. "Among them was the story of the *Stiefmutterchen* (German for Little Stepmother, better known as the pansy). He would pick a pansy, and handling it with care so as not to injure its delicate beauty, proceed to introduce us to the little stepmother and her family":

"Today our father is 95 years old. He can no longer see the beauty of a flower. Recently we asked him if he would like to relive his life. 'Oh yes,' was the answer, 'if I could again be a gardener.'"

The Stiefmutterchen

The stepmother is number 1. You will note that she is wearing the most beautiful dress. Numbers 2 and 3 are her own daughters. They, too have lovely dresses, although not as beautiful as her own. Numbers 4 and 5 are her stepdaughters. They wear plain, unadorned dresses. If you turn the flower over, you will see the sepals which represent chairs. The stepmother has two chairs while her daughters, numbers 2 and 3, each have a chair to sit on. But sadly, her stepdaughters, numbers 4 and 5, must share one chair.

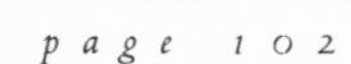

Poppies

Abby had been sick in bed for over a week. Cranky, aching, and tired of being indoors, she begged me to take her outside for a walk. "Sorry, Abby," I said, "Mom says you can't go outside until you are 100% well." "Oh, please, Sharon, please let's do something different. I am so tired of looking at the same thing every day, all day long!" "You are going to do something very different today," I replied. "You're going to watch a flower being born." I reached behind the door, pulled out a terra cotta pot filled with Iceland poppies, and set it on a sunny windowsill.

In the silent language of flowers the poppy is the flower of CONSOLATION.

We read the morning away, and as we leafed through page after page of *The Secret Garden*, we watched a plump, hairy poppy bud slowly split, showing traces of orange along the seams. Then, quietly, almost sneaking by us, the pod fell away. During the next two hours we watched as the papery, orange petals unfurled and the wrinkles disappeared. I think that I was as excited and as touched by the poppy's birth as Abby was. I left the pot of poppies with her and every day she called me to give me a "poppy progress report".

Dandelions

Millie Heath told me, "Every child knows how to tell the time by a dandelion clock. You blow as hard as you can, and you count each of the puffs left. An hour to a puff." (Dandelion clocks tell fairy time.)

Blow upon a dandelion,
Close thy eyes,
Chant
This year,
Next year,
Sometime,
Never,
If one clock remains,
We will be together.

OLD SONG

THE YOUNG DANDELION

I am a bold fellow
As ever was seen,
With my shield of yellow,
In the grass green.

You may unroot me
From field and from lane,
Trample me, cull me—
I spring up again.

I never flinch, sir,
Wherever I dwell;
Give me an inch, sir,
I'll soon take an ell.

Drive me from garden
In anger and pride,
I'll thrive and harden
By the road-side.

Dinah Mulock Craike

Dandelion Fortune Telling

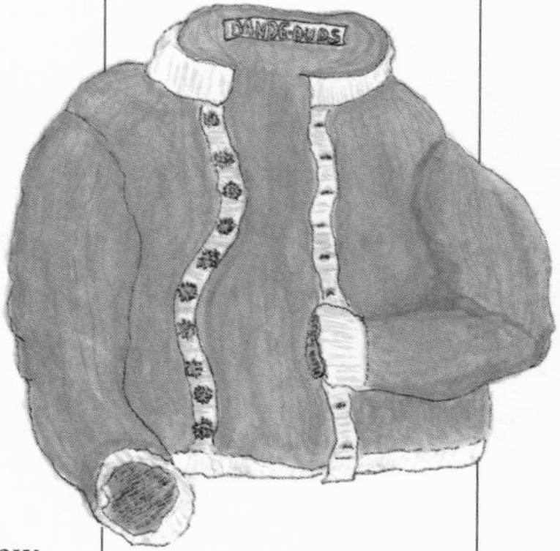

It has been said that each of the tiny, feathery "clocks" on a dandelion has the power to tell time, divine emotions, and sail secret, winged messages to a loved one's soul.

Boys and girls of a hundred years ago picked "old man" dandelions, turned toward the direction of their faraway love, and blew once. If a single feathery seed remained, they knew they were not forgotten.

When children wanted to find out how many children they would someday have, they would pick a dandelion puff, close their eyes, and blow. If ten seeds remained, it foretold a large family!

From *Dandelion Clocks and Other Tales* by Julia Horatia Ewing London, 1894.

OVER THE CLIMBING MEADOWS
WHERE SWALLOW SHADOWS FLOAT,
THESE ARE THE SMALL GOLD BUTTONS
ON EARTH'S GREEN WINDY COAT.

FRANCES FROST

Sweet Lavender

Lavender hedges often were planted around cottages in England. It was believed that they offered protection from evil....

There's nothing else like the scent of sweet lavender—fresh and clean straight from the garden in summer, or hauntingly nostalgic among the linens in winter. Lavender wands are so simple for children to make, and such a welcome gift for grandma or an older friend. Here's a story about lavender wands told to me years ago by a customer at Heart's Ease.

"Do you know I still have a lavender wand—sometimes we called them lavender cages—that I made over fifty years ago?" she said. "I keep it in a drawer and when I run across it I feel a whole flood of garden memories washing over me.

"Grandma would take us out into her cottage garden in the early morning as she cut her flowers and herbs for the day. I can still feel the warmth of the sun and I swear I can smell the lavender and the pinks. We would pick a handful of lavender and tie the heads together. Then, we would bend the stems back over the tied heads and tie them together—it looked like a little bottle. In fact, I think my neighbor called them bottles.

"Grandma would give us bits of ribbon from her sewing basket and we loved to weave the ribbon over

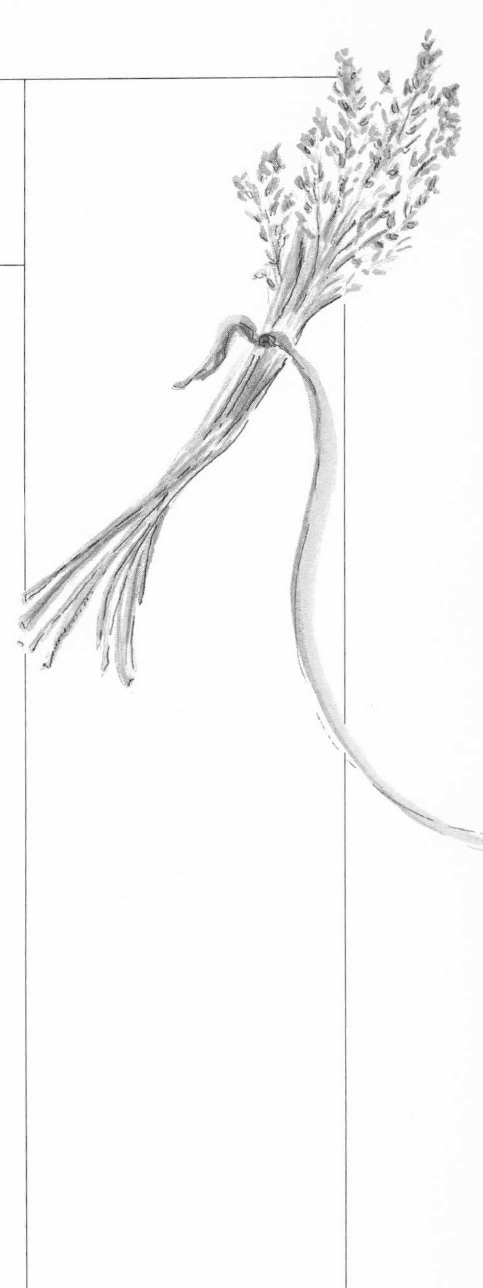

SWEET LAVENDER!

I love thy flower
Of meek and
modest blue,
Which meets the morn and
evening hour,
The storm, the sunshine,
and the shower,
And changeth not
its hue.

Agnes Strickland

and under the lavender stems. After the lavender wands were thoroughly dry we wrapped them in tissue paper and were so proud to give them to our schoolmistress the first day of school. We always had plenty to give as Christmas gifts too."

To make a lavender wand, pick an odd number of stalks in full bud, just before the flowers start to open all the way. Thirteen or fifteen stalks makes a nice fat wand. Bundle the stalks together and tie them firmly just below where the buds start. Holding the bundle with the stems pointing upward, carefully bend each stem down so that the buds are enclosed in a little "cage" of stems. Tie the stems together with ribbon just below the enclosed cluster of buds, or use a long piece of ribbon to weave over and under the stems, making a little basket to contain the buds.

Touch-Me-Not

During the past years the "grown-up kids" I have interviewed have often mentioned jewelweed. This plant seems to be one that tickles the dickens in all of us. I had my first experience with jewelweed at Audubon Camp in Maine. We were hiking the Roger Tory Peterson Bird Trail skirting a large pond. One of the men on the hike asked me if I liked jewelweed. I told him that I had never run into it and would love to look at it with my magnifier.

He pulled me over to a bush covered with beautiful yellow-spurred flowers spotted with red. Hanging from a fragile stem, swaying gently in the breeze, it was easy to see why these would be called jewelweed. Crouching over a blossom, not touching it, I looked up and asked why it was called "touch-me-not", too. My friend grinned from ear to ear, stretched his hand out to touch the flower, and ZING...like a firecracker exploding, seeds flew in all directions.

My question was answered and although I don't like to admit it, I did enjoy touching the "snapweed" all the rest of our hike. Sure feels good acting like an irrepressible kid again! I might mention that at least a dozen other "adults" on our hike were doing the same thing.

Eardrops

Eardrops of gold with

red rubies beset,

Hang from the ears of a

dear little maid.

"Where did you get them,

my darling, my pet?"

"Down by the brook you

can pick them," she said.

Margaret Morley

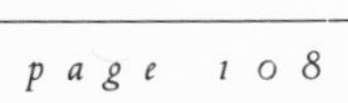

Violets

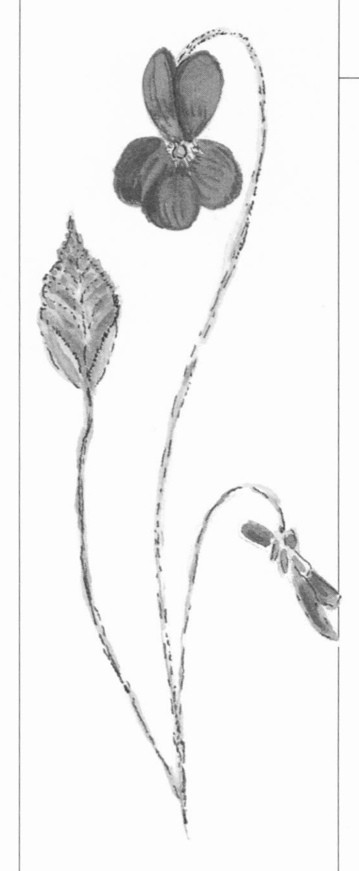

Who bends a knee
Where violets grow
A hundred secret
Things shall know

You would think that the lovely, fragile violet would be the last thing a young boy would use to have a battle, wouldn't you? It was with a reluctant head shake and grin that Thomas Stanley related to me the violet tug-of-wars he fought in Shelbyville, Kentucky in the early 1900s.

"We would go out to the woods and pick a bunch of violets," he said. "You know that little hook they have on them where the petals are? Well, we would try to hook each other and see who could snag a violet head and rip it off. Whoever amassed the most violet heads was the winner of the violet tournament."

Origin of Violets

I know, blue modest violets,
Gleaming with dew at morn—
I know the place you come from
And the way that you are born!

When God cut holes in Heaven,
The holes the stars look through,
He let the scraps fall down to earth,—
The little scraps are you.

—

Anonymous

Bleeding Hearts

Every spring, at the mossy base of my old sundial, a small miracle quietly unfolds. First, slender spears of green, then the heart-shaped leaves of old-fashioned violets appear. Amongst the violets another leaf shape emerges almost unnoticed. And one day, without fanfare, a fragile wand of parading bleeding hearts (*Dicentra spectabilis*) sways gently above my violets.

In my garden I can always calculate the degree of child left in a person by how they respond to my bleeding hearts. The true measure is the excited cry of joy and recognition when they notice the hearts. Then, the almost-instantaneous kneeling and close examination of the dangling flowers. I find that the only difference between the child of eight and the child of eighty is how long it takes to rise from that kneeling position!

Bleeding hearts are one of the most loved of the old-fashioned garden flowers. So many stories have been told to me about how the flowers were hung over ears as earrings, woven into hair as a dancing heart tiara, or dissected to reveal a lady-in-a-bathtub or a man-in-a-gondola.

My favorite is the story of the beautiful Princess

The true measure is the excited cry of joy and recognition when they notice the hearts.

A browned, shadowy silhouette of a pressed stem of bleeding hearts marks this poem in the pages of my 1884 St. Nicholas magazine.

"I know where there's a beautiful shoe—
Tiny and fair and ready for you;
It hides away in the balsam flower,
But I'll find you a pair in less than an hour."

"Thank you my laddie; now this I'll do,
I'll pluck a heart-flower just for you.

The hearts hang close on a

bending spray,

And every heart hides

a lyre away.

"How shall you find it? I'll

tell you true:

You gently sunder the heart

in two,

And, under the color,

as white as milk,

You'll find the lyre with its

strings of silk."

So now, child-at-heart, you see that one man's princess is another man's fairy-sized lyre.

Dicentra who wandered away from her walled garden and became lost in the darkness of an ancient forest. The princess fell prey to an evil crone angered to have her privacy disturbed. In an instant, Di was reduced to a fraction of her normal size and entrapped in the satiny pink folds of an oddly- shaped flower.

The old crone cackled happily and told Di that she was to remain forever imprisoned unless discovered and released by an innocent youngling.

Little did the crone know how tempting the Princess would be to any passing child! Only three days passed before a party of riders stopped for water in the forest. Drinking from a stream on bended knee, a boy glanced up, spied the dancing wand tipped by a pink and white heart, and plucked it (as innocent children will do). Short, plump fingers folded back Di's voluminous pink skirt, and the lovely princess was saved!

Dicentra is the perfect pressed flower. Picked early in the morning (after the dew has dried) and tucked between the pages of a thick dictionary, bible, or phone book, the flower is soon transformed into a perfect flat, papery little heart. After a few weeks of drying the bleeding hearts can be glued on greeting cards, valentines, or placecards for a special garden party.

If you gently fold back the hoop skirt of the Bleeding Heart, inside you will find Princess "Di" Centra

Screechers

Millie Heckman Huffaker was raised on a ranch in Exeter, Tulare County, California, in the early 1900s. "We always had to make our own amusements and toys," she said, "and we always seemed to have plenty to do. The toys that were my favorites were my grass screechers.

"You pick a piece of Johnson grass and hold it between your thumb and first finger on each hand...like so. Pull it tight, hold it on its edge, and blow." Millie blew, and grinned from ear to ear; it was the first time she had tried it in sixty years. "You can make different sounds and screeches and whistles, depending how you breathe and blow. You can really get good at it and play tunes and fool people."

AFTERNOON ON A HILL

I will be the gladdest thing
Under the sun!
I will touch a hundred flowers
And not pick one.

I will look at cliffs and clouds
With quiet eyes,
Watch the wind bow down the grass,
And the grass rise.

EDNA ST. VINCENT MILLAY

Hey, I've found some moneywort.
Some day I'll be rich!
Or I wonder if it's checkerberry?
I don't know which is which.

Look, don't touch that blade of grass,
Just keep away from it!
For see that frothy, bubbly ball?
That's snake spit!

Cover your lips,
The darning needle
Loves to sew 'em up!
Who likes butter?
Lift your chin-
Here's a buttercup.

"D'ye ever whistle a blade of grass?
Look, I got a fat one...
You slit it, see?
Here's one for you—
There's no snake spit on that one."

Witter Bynner,
Child Life Magazine, 1937

 *the fluttering and*

the pattering of

The green things growing,

How they talk each to each

When none of us are

knowing.

Dinah Mulock Craike

How To Make A Whistle

FIRST TAKE a willow bough,
Smooth and round and dark,
And cut a little ring
Just through the outside bark.

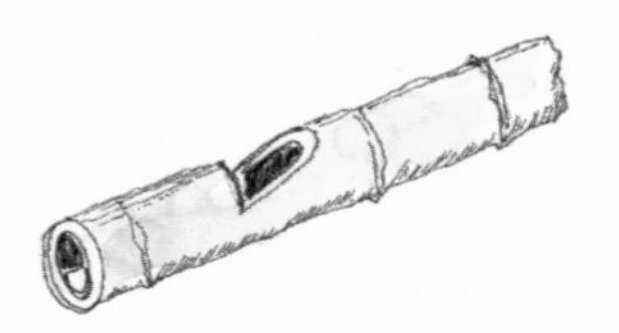

THEN TAP and rap it gently
With many a tap and pound,
To loosen up the bark,
So it may turn round.

SLIP THE bark off carefully,
So that it will not break,
And cut away the inside part.
And then a mouthpiece make.

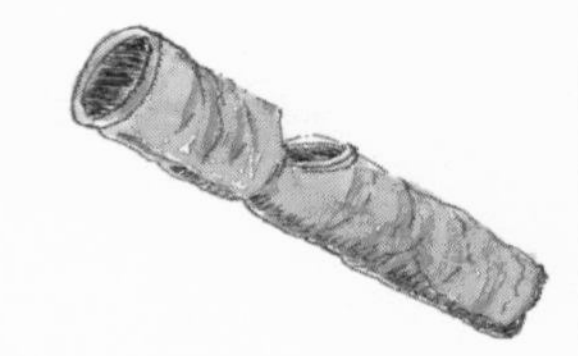

NOW PUT the bark all nicely back,
And in a single minute
Just put it to your lips,
And blow the whistle in it!

ANONYMOUS

About Walnuts

John Arnold of Kokomo, Indiana, was famous for his walnut sailing vessels back in the 1940s. He'd fill a halved walnut with sap, glue, mud, or a gumdrop and stick in a toothpick, twig, or matchstick for a mast. Either a leaf or piece of paper would suffice for a sail, and he would spend hours sailing his fleet in Wildcat Creek, any rain-swollen gutter, or even Gram's wash-tub. When John had a son, he found himself constructing walnut boats again and delighting not only his son and his young friends, but himself as well!

Sailing off to

unknown seas

Walnut boats

with sails of

leaves

I once heard a story about a little girl who was very poor. Her birthday arrived and her family had no money to buy the sugar, flour, and eggs to bake her a cake. Missing out on a cake didn't bother her, but she was heartbroken about not having candles to wish upon.

When suppertime arrived, the girl swallowed the lump in her throat and resolved not to cry. Dinner was very meager, white bread with lard, poke greens, and a pot of navy beans...without ham. The little girl bravely ate her dinner and bantered cheerfully with her brothers and sisters.

After the table was cleared, just about the time the cake should have been served, the lights went out without warning. Mother, father, and all of the children paraded into the room holding the huge old crockery mixing bowl filled with water and a flotilla of walnut boats. Inside each walnut boat (one for each year of her life) was a small birthday candle glowing brightly. The little girl closed her eyes tightly, made a wish, and blew so hard the little boats scudded around the edges of the bowl as each and every candle was extinguished.

Walnut Baskets

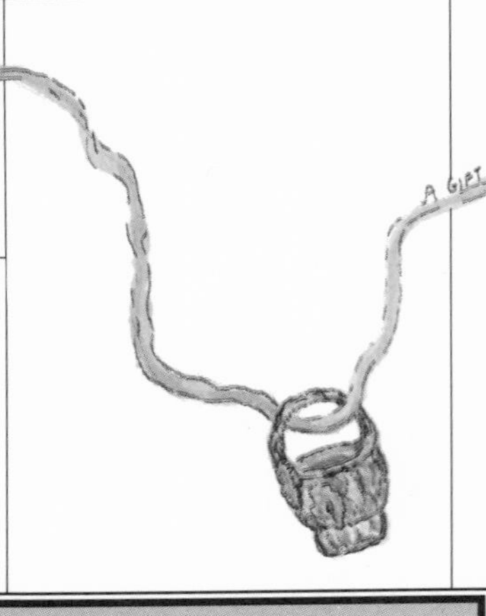

Sometimes the simplest of gifts can give the most joy. My friends David, Julee, and Summer Krause stopped by on Christmas Eve and delivered a tray of cookies and a beautifully decorated golden box. I couldn't imagine what was inside. Jewelry? Really, there was nothing I needed.

On Christmas morning I eagerly unwrapped the gift from the Krause family and let out a shriek of excitement. Inside the box was a set of six tiny walnut baskets. Each basket was carved by David, and then Julee and Summer trimmed them with miniature tablecloths and tied the handles with small red bows. It looked like a perfect picnic ensemble for a family of mice.

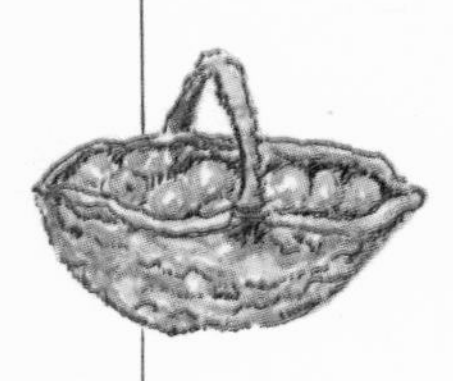

Now, every year I fill the baskets with pearly pink berries and hang them from our Christmas tree.

Today my dear friend Julie Whitmore brought my birthday gift—a tiny, tiny peach pit basket on an old and faded pink satin ribbon. Obviously, this was once somebody's prize necklace. I have set it next to my walnut baskets and it is positively dwarfed by them. Someone lovingly sanded, carved, and polished this little gem. I only wish I knew its story: from what garden, in what state, and carved by whom? I shall treasure it always!

Acorns

Who could help being intrigued by acorns in their little caps? I still enjoy walking through the woods and stuffing my coat pockets full of them!

The Battle of the Acorn Tops

Pick freshly fallen acorns and run a toothpick or nail through the acorn cap and halfway into the acorn (you must pre-drill a hole). The acorn top game may be played alone or with a friend.

First you empty a tabletop and set some of your acorn tops spinning. They may bump into each other and spin off of each other, but this is part of the battle. If one spins off of the table he is the loser. Whoever spins longest on the table is the winner. Whole armies of acorn tops may be made and set to spinning at the same time. Flowers of a certain color can be impaled on each tip. Perhaps red geraniums could top one army and blue cornflowers top another. Whichever army has the most spinners at the end of the battle wins. It is a sight to see all of the tops wildly spinning with vivid colors!

Gourds, Gourds And More Gourds

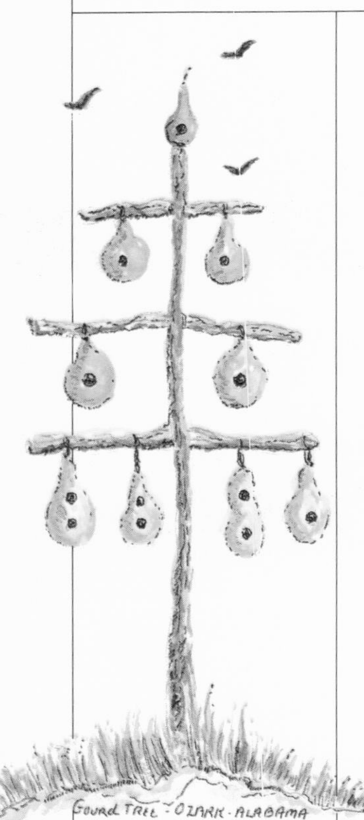

This story was told to me by Dorothy Greeman Peterson, who grew up in Turner County, Georgia, on a 200-acre peanut farm.

"In the springtime we would make houses out of gourds for the martins, sparrows, and other birds to lay their eggs in and raise their young. It was a beautiful sight to watch those birds flutter around the gourd houses and build nests. And they made such beautiful music!

"To make gourd houses you take green gourds, cut holes in them the shape of windows and doors, big enough for a bird to get through, scrape out the seeds and pulp, and hang them up to dry. We would poke holes through the tops and run wires through so we could hang them from a gourd tree after they dried.

"It was so much fun to make the gourd houses and watch the goings on. In the South gourds grow wild, usually among the watermelons or on the side of the road. We used leftover gourds as dippers on the back porch: just cut the gourd in half and dig out the pulp and seed and let it dry out. Oh, but the birdhouses were my favorite."

IT WAS A BEAUTIFUL SIGHT TO WATCH THOSE BIRDS FLUTTER AROUND THE GOURD HOUSES AND BUILD NESTS.

Here is a tip from an 1850s seed catalogue: "If, after a few gourds have set, the ends are pinched off the vines, the gourds will grow larger and better. Harvest the fruits before frost and when fully ripe, handling carefully to avoid bruising. Then, with a sharp knife slice away the unwanted parts to form the container, dipper, or whatever. Remove seeds, wash the gourd with a strong household disinfectant, and put it in a dry and ventilated place. Turn it every few days while drying, which will take weeks and in some cases months. You can polish it with paste wax."

•

YOU CAN

b a s k e t s

MAKE LOTS

b i r d h o u s e s

OF GREAT

s p o o n s

THINGS

r a t t l e s

USING

t o y s

GOURDS

•

If you haven't tried raising gourds you are missing a wonderful experience. Gourds provide a fragrant, fast growing shade screen, and the flowers are lovely. At Lewis Mountain Herbs in Ohio, a corner of the garden has a simple wooden framework totally covered with gourd vines and hung gaily with huge, ripening gourds. The gourds are harvested and dried for use as planters, birdhouses and other things, but during the summer the vines provide a delightful shady spot.

An old Burpee catalog for 1887 states, "Sugar trough gourds are useful for many household purposes such as buckets, baskets, nest-boxes, soap and salt dishes. They grow to hold from four to ten gallons each, have thick, hard shells, very light but durable, having been kept in use as long as ten years." Dippers and birdhouses were made from the dipper gourd, spoons were made from the spoon gourd (or it was left whole to form a darning egg for stockings or to become a rattle for the baby because of the dried seeds inside). Dippers and spoons made from gourds had the advantage of handles that don't get hot.

"A wonderful gourd toy that was often made in the late 1800s was a gourd man. He was simply stuck with twigs for arms and legs and a little face was painted or scratched on the top. But my favorite gourd toys were the coach gourds. An elongated gourd would have one quarter of the narrow part cut off so that it then resembled a coach shape. The pulp would be cleaned out and a long twig pushed through for an axle for the coach's wheels. The wheels could be made of thin log rounds, sliced buckeyes, or slices of cucumber.

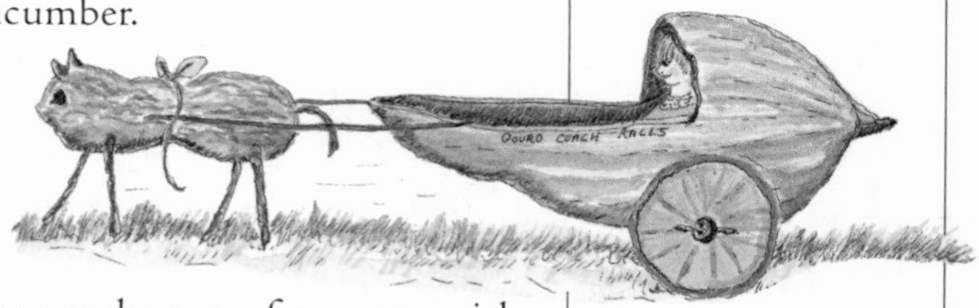

The horses were easy to make out of peanuts with twig legs and drawn-on faces. Long strands of grass were tied around the peanuts to hold the harness in place. Driving the coach could be a lovely flower lady or a peanut lady or even a corncob lady. Whoever drove it, the ride would be fun!

In the 1880s, a favorite flower lady doll was made of gourd flowers! Harvest the tiny gourd just as it begins to form. This will be the doll's head. A gourd leaf is used as her clothing, and a hat can be made from another leaf. Twigs or toothpicks can be used to hold all of the parts of this lady together.

Betting On Hops

On a hot, clear, spring afternoon I was sitting on Georgie Van de Kamp's veranda with a group of gardening friends. We were discussing gardens and things we remembered about gardens and Carol Bateman piped up, "We used to place bets on hops!"

> HOP VINES GROW SO ❖ FAST YOU CAN ALMOST ❖ SEE THEM CLIMBING ❖

"Bets on hops?" I asked, "What in the world do you mean?" Carol laughed and said that hops vines grow so fast you can almost see them climbing.

"During our family's summers in Yosemite we would place bets on how much the hop vine would grow during the night," she said. "In the morning, we would go out and measure and do you know that some nights it would actually grow 6 or 7 inches!"

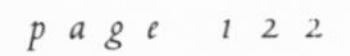

A Mouthful Of Hearts

Thomas Stanley was in his late 70s when he told me the story of how he and his friends turned grass into hearts and fooled the girls.

"We would pick a bunch of shepherd's purse—you know how it has the flat seed pods that look like little hearts? Well, we would hide the hearts under out tongues. Then we would stand in front of the girls and pick a handful of grass and say, 'We can chew grass and spit out hearts.'

"The girls would always laugh at us and tell us we couldn't, whereupon we would gobble up the blades of grass and spit out the shepherd's purse hearts. Got 'em every time!"

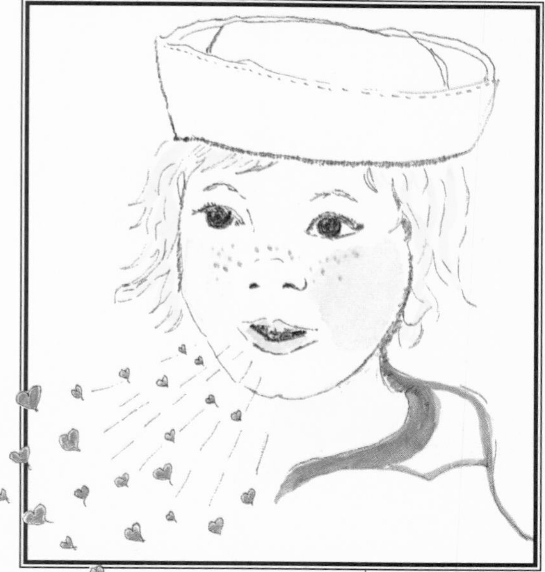

Trumpet Flower Bubbles

"we spent all of our time on the huge front porch which was draped and shaded by an old trumpet vine."

Margaret Crittenden Sparks told me this story about growing up in Topeka, Kansas, in the 1920s.

"Grandma always said that summers in Topeka were so hot and the air so thick and heavy that the wheat could stand up on its own after it was cut.

"Playing out in the sunshine or in the house was unbearable. It seemed as though we spent all of our time on the huge front porch which was draped and shaded by an old trumpet vine.

"My sisters and brothers and I had mixed up a bucket of soap flakes and water. We were planning to spend the day blowing bubbles, but much to our dismay we could find only one bubble pipe.

"For some reason, our eyes all seemed to settle on the red trumpet flowers at the same time. We each picked one, pulled out the stamens to create a little hole, pinched the hole closed, poured in the soapy water, tilted the blossom up, and blew from the wide part of the petals. It worked perfectly! We never argued over the bubble pipe again—we had an unlimited supply!"

The Magical Pumpkin Patch

A fairy seed I planted,
so dry and white and old -
There sprang a vine enchanted,
with magic flowers of gold.

I watched it,
I tended it, and
truly by and by
It bore a Jack-o-Lantern
and a great
Thanksgiving Pie.

Marjorie Barrows
Child Life Magazine
1937

This story was related to me in the early 1970s. Mammy is no longer living, but her story is as delightful today as when I first heard it.

"When we were little kids in Alabama we just didn't have any money or any toys like kids have now. We had to make all our own fun and mischief—and we did! Let me tell you the pumpkin patch story.

"The kids in my family were pranksters. We could stir up a storm without a cloud in the sky. We would sneak out on moonlit nights and head straight for old John Henry's watermelon and pumpkin patch. Sometimes we would just cut the heart out of the biggest melons in the patch and gorge ourselves on that juicy, sweet heart and then we would giggle and roll around so stuffed and proud of ourselves. Isn't this a shameful tale? Anyhow, we loved to pull his biggest pumpkins out of the patch and roll them down the hills. Sometimes we would build pumpkin walls from one side of the road to the other. We never, ever thought that John Henry would dream we were the culprits.

"One dark October night we stole over the fences and through the kudzu thickets into John's pumpkin patch. We went out into the middle and headed for the biggest pumpkin there. My brother Asa rolled that huge pumpkin over and there, in huge letters carved right in the pumpkin were the words...'THOU SHALT NOT STEAL'! We were right frightened! We turned over another one...'HE SEES ALL'! At this we looked up and around, let out screams, and high-tailed it back through the kudzu and home faster than we'd ever travelled before.

"Well, now I'm older and wiser. I know that old John had us figured out pretty early on. Finally, fed up with all our pranks, he had picked out the fairest dozen pumpkins in his patch and had scratched warnings into their soft skins as they were growing. As they grew so did the warnings and by the time we saw them they were darn near as big as billboards.

"Raising my own family, I always let the kids plant a pumpkin patch of their own. They are so comely and lovable, those huge old pumpkins. I love to scratch my grandchildren's names and their birthdates and sometimes phrases into the pumpkins as they are just starting to put on their growth. The kids love them, and they are always so tickled to have their own namesake pumpkin!"

Why don't you do this for your children? Give them their own magical pumpkin, named especially for them, and watch their wonder and joy. Recently, I saw a pumpkin patch where a farmer had customized his crop. He took orders in the spring and actually scratched names and addresses or phrases on the pumpkin. For example: WELCOME FRIENDS! These pumpkins last a long time; you could have one sitting on your front porch or doorstep all through the autumn and sometimes even into late spring.

Grass Slides

"My heart

would race as I

made the

season-opening,

bumpy ride

down the hill."

I grew up in a tiny valley surrounded by high, grass-covered hills. Every spring, when the grass got about up to our waists, an excited wave of anticipation swept through the valley. Time for the grass slides—time to start collecting cardboard boxes. Everyone's trash became fair game. Early on Saturday morning we would rally together with boxes of all sizes. We tore the boxes down, stamped them flat, and then the fun began.

The ascent up the green hillside was slippery and riotous. When we reached the top and looked down, our houses were toy-sized and the people working in their yards looked like bugs. The first ride down was the roughest. We drew straws to determine who would make the "rough run". I usually lost; in fact, I think I always lost. Could it be because I was the only girl in the gang? My heart would race as I made the season-opening, bumpy ride down the hill. Our whoops and hollers could be heard from one end of the valley to the other as our spring-time sledding party slipped through the sweet, green day.

Yesterday I was talking to some friends about grass slides. Kim Cory remembered racing down Laguna hillsides over a carpet of pungent, wickedly slick eucalyptus leaves. Carolyn Germain described wild runs down Plymouth Avenue in St. Francis Woods in San Francisco in the early 1950s. "And those hills were *really* terrifyingly steep," she said. Jeff Prostovich remembered that when spring arrived and sliding began, he had to find numerous excuses for arriving home late with grass stained pants. I wonder if any of us would have the courage to do it now?

Dragonflies

When I was a youngster maybe four or five years old, I was terrified to run along the hollyhock trail to Grandmother's house. I would peek out of our screen door, check to make sure the coast was clear, and then streak through the garden to Grandmother's back porch. One day I nearly bowled her over as I leapt up the stairs and into her kitchen.

"What on earth is scaring you, dear? she asked. "The dragon monsters," I replied, with my head bowed in shame. "Do you mean our friend the dragonfly?" she asked. I nodded, ashamed to have finally confessed my biggest fear. "Let me tell you about dragonflies, my dear," she said. "They are great hunters and while you are playing outside in the garden they are your guardian angels. They patrol the sky above you and snatch insects out of the air before they can bite you or bother you. Indians believed that the dragonflies, which are sometimes called darners, passed back and forth between the water and the air and stitched the rain clouds into the sky. When it rains it is because the dragonflies

This is a DAMSELFLY. They are smaller than dragonflies and when they rest, they fold back their wings - which is something dragonflies don't do.

have decided to rest for a day and they don't attend to their duties of cloud-darning. On their day of rest the clouds loosen up and drop their rain on the dry earth, awakening sleeping seeds and giving birth to all of the beautiful flowers in our garden. Do you remember the dragonflies ever being out on a rainy day?" I puzzled over that question for a minute and then had to admit that I didn't.

"So, they're my guardian angels and they're protecting me from insects and they put clouds in the sky and they help us have flowers in our garden," I stated bravely. "Then I won't ever be afraid of them again." I turned, walked slowly down the steps, past the iris bed and straight over to the platoon of hollyhocks on the pathway. A pair of jewel-colored dragonflies clattered noisily into each other defending their territories. My heart jumped, but only once, as I turned and waved to Grandmother.

A living rush of light he flew

TENNYSON

Riddles

This flower riddle was given to me by Margie de Lyser of Cambria, California. The unsolved riddle was found folded up in a box of old letters hidden in a trunk. Can you match the clues with the flowers named here? Answers are on page 144.

JONQUIL

FLAG

MAIDENHAIR FERN

POPPY

CARNATION

TIGER LILY

LAMB'S QUARTERS OR PHLOX

JACK-IN-THE-PULPIT

FOUR-O'CLOCKS

LILAC

ANGEL'S TRUMPET

GRANDMOTHER'S GARDEN

In Grandmother's garden strange plants you will see,
And if you guess rightly you'll find twenty-three.
They are all out of order for climate and time,
And arranged in this manner to give the words rhyme.
Just inside the gateway some clergymen stand (1)

With a bugler who plays in the heavenly band (2).

The name of a boy and an old-fashioned weapon (3),

You will find with the cares of all single men (4)

In Grandmother's garden we likewise behold,
Some plants that remind us of sheep in the fold (5).

And near them all standing, too stately to bend,
That which the soldier has died to defend (6).

A state in the South and a one-year-old child (7),

Form a beautiful background in this garden wild.
Here too, with its head held haughty and high,
The dread of the jungle lurking near by (8).

Yet farther, a fairy wand all made of gold (9),

And the pride of the mermaid so fabled of old (10).

A little white sin and a spinster's pet charm (11),

In yon shady thicket is sheltered from harm.
A time of the day (12) and a little frog's walk (13),

And a part of the face we use when we talk (14),

The child of a suffragette known in our land,
With one letter changed to good spelling command (15).

A pet name for father (16) and an embrace so sweet (17),

Are all to be found in this quiet retreat.
But ah! Here a beauty so perfect to see—
The serf of a Mexican followed by "e" (18).

The hope of our Pilgrims (19), an attempt made to bite,

And a hideous monster once slain by a knight (20),

A mode of conveyance, a word meaning tribe,
Attracts our attention, and is Grandma's pride (21).

A shot from a cannon, and part of the foot (22),

While along the rough pathway Grandma has put
A pleasant expression, and one sharp-edged tool (23).

Now please try to guess them and stick to the rule.

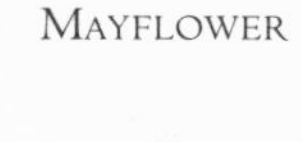

MAYFLOWER

TULIPS

HOPS

CRESS

PEONY

SUN ROSE

GOLDENROD

BACHELOR BUTTON

VIRGINIA CREEPER

SNAPDRAGON

SMILAX

MISTLETOE

London pride
Lad's love
Black-eyed Susan
Jacob's ladder
Bishop's hat
Blush rose
Speedwell
Eyebright
Fair maids of France
Poppy
Love-in-a-mist
Sweet William
Wake robin
Thrift
Marguerite
Johnny jump-up
Maiden pink
Bleeding heart
Coxcomb
Wallflower
Rambler
Bachelor's buttons
Ladies tresses
English daisy
Four o'clock
Lady's slippers
Spring beauty
Goldenrod
Honesty
Tulips

LOVE AMONG THE ROSES

This was sent to me by Julie Whitmore. She found it in an old issue of *Modern Priscilla Magazine* dated 1928. Answers are on page 144.

Yellow was especially becoming to little (1), and so when (2) that dashing (3), invited her to a party at (4), she gratefully accepted this proof of the (5), and put on her yellow dress and yellow (6) in honor of the occasion. First, she carefully arranged her (7), and then tiptoed softly out of the house so as not to (8), her little brother. The mirror in the hall showed her that she was a (9), and that if her name had only been (10), she would have been a real (11). Her escort's (12) leaped high as he saw her, though, not to be outdone, he had with careful (13) polished his own (14) until they shone like a (15). "Not one of the (16) can equal your appearance!" he exclaimed proudly. "England forever!" A tinge of (17) showed on her face as he spoke with such (18), for behind it she read aright his (19). But she only answered him demurely, "I hope I shall not be a (20)." "Far from it," he answered warmly. " I would scale (21) itself for a dance from you." By that time they were at the party. "(22) exclaimed her escort to a boy at the door, "and give her your seat!" "Never," answered the young (23) disagreeably, and when pressed, he gave her lover such a blow that he saw his (24). But when he saw the (25) approaching he ran away. "Oh, Billy, are you hurt?" she sobbed wildly. He opened his (26) with love and answered feebly "Will you be mine?" "Ask (27)," she answered shyly; while a (28) to her cheek. His (29) answered in the old, old way and all we can do is to wish them (30).

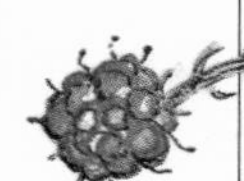

As soft as silk,
As white as milk,
As bitter as gall,
I'm rather tall,
And a green coat
covers me all!

Answer - The milkweed pod

Ten Curious Berries

There's a berry which makes my pony's bed;

And another one which is green when red;

And there's one which rubs you all the wrong way;

And another which swims and quacks all day;

There's one you can play, to beguile your care;

And one at their necks the ladies wear;

There's a berry which seems to be much depressed;

And one is a bird with a speckled breast;

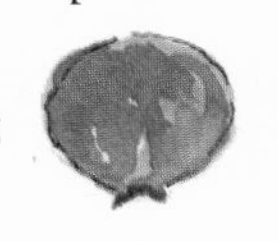

There's one we can see when the tide is low,

And the last you will be when you older grow.

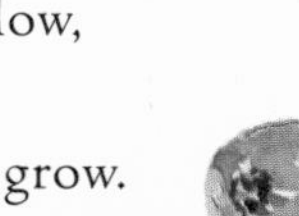

Answers: Strawberry, Blackberry, Raspberry, Gooseberry, Checkerberry, Mulberry, Blueberry, Partridge-berry, Barberry, Elderberry

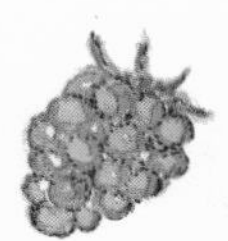

Some fill me,
Some beat me,
Some kill me,
Some eat me;
I creep and I fly,
And my color is green;
And though I'm a season
There's quite a good reason
Why my end or beginning there's no man hath seen.

Answer - Thyme

Epilogue

I am young again and sitting in the cool darkness of Grandmother's early California bungalow. The smells of oatmeal cookies and simmering marmalade are wrapped around me like warm hands.

Outside, hummingbirds are dipping into the brilliant, red bottlebrush as Grandmother grabs her old straw hat and looks over at me. I wait for the familiar words, "Let's get busy, Sharon, time to go outside and see just what's happening in our garden."

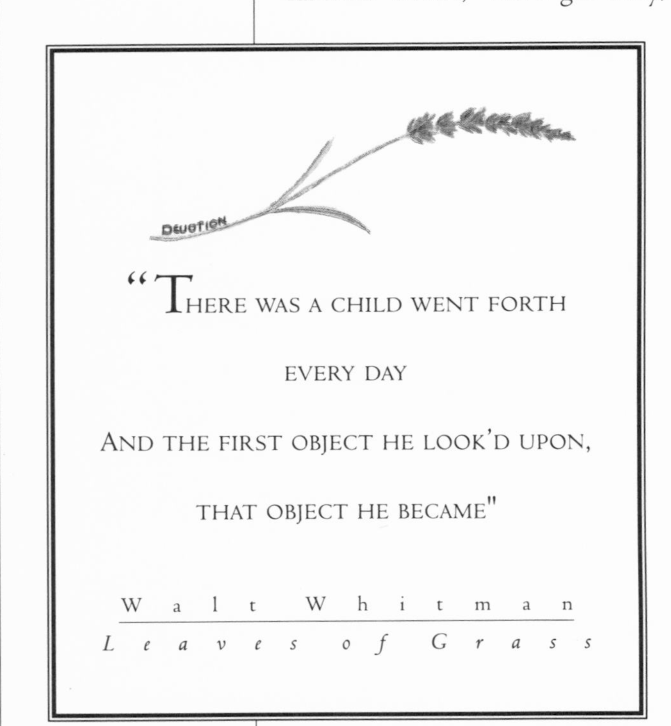

"THERE WAS A CHILD WENT FORTH
EVERY DAY
AND THE FIRST OBJECT HE LOOK'D UPON,
THAT OBJECT HE BECAME"

Walt Whitman
Leaves of Grass

My heart still soars when I smell oatmeal cookies, freshly turned spring earth, carnations in full sunshine. I relive over and over the joys and surprises of each day in our garden. I can sit quietly and string those sweet garden thoughts together, memory upon memory, like my summer garlands of tiny, pink rosebuds.

I know now that the gift my Grandmother Lovejoy gave me is an ancient one that runs like a tenacious woodbine through the childhoods of the hundreds of gardeners who have contributed to my book. A gift which has been quietly passed on from loving aunts and uncles, neighborhood friends, mothers, fathers, grandmothers and grandfathers, teachers, naturalists—it is the magical gift of sharing, a sharing of the reverence for the

earth and of the simple miracles of each unfolding day. Sharing the faces in a pansy, the opening of a poppy blossom, the taste of fresh hollyhock cheeses, or a pumpkin big as Cinderella's coach—small, seemingly insignificant sharings that will pop into our children's minds when they smell a familiar flower, or watch the sunflower's slow dance through the day. And they will say to their children, "Come over here, did you ever play tops with an acorn? Want to make a day-long jump rope?" And the traditions and love will keep lengthening, like my never-ending chains of summer rosebuds.

F A R E T H E E W E L L !

Let our children look upon the flowers of the garden.....

Bibliography

Blanchard and Lea. *The Handbook of the Sentiment of Flowers.* Philadelphia: 1847.

Bailey, Liberty Hyde, and Ethel Zoe. *Hortus Third.* New York: Macmillan Publishing, 1978.

Barrett, E.L. *The Doll's Own House.* Springfield, Ohio: G.C. Hall and Company, 1882.

Barrows, Marjorie. *One Hundred Best Poems for Boys and Girls.* Racine, Wisconsin: Whitman Publishing Co., 1930.

Beard, Adelia, and Lina Beard. *The American Girl's Handy Book.* New York: Charles Scribner's Sons, 1887.

Bralliar, Floyd. *Knowing Insects Through Stories.* New York: Funk and Wagnalls Company, 1918.

Brickell, Christopher, and John Elsley. *The American Horticultural Society Encyclopedia of Garden Plants.* New York, Macmillan Publishing, 1989.

Burgess, Thornton. *The Burgess Flower Book for Children.* Boston: Little, Brown and Co., 1923.

Bynner, Witter. "Read Aloud Time". *Child Life Magazine,* May, 1937.

Crane, Walter T. *Flowers from Shakespeare's Garden.* London: Cassell and Co., 1906.

________________. *Flora's Feast.* London: Cassell and Co., 1899.

Curtis, Mary I. *Stories in Trees.* New York: Lyons and Carnahan, 1925.

Earle, Alice Morse. *Child Life in Colonial Days.* New York: Macmillan Publishing, 1899.

________________. *Home Life in Colonial Days.* New York: Macmillan Publishing, 1900.

________________. *Old Time Gardens.* New York: Macmillan Publishing, 1901.

________________. *Sun Dials and Roses of Yesterday.* New York: Macmillan Publishing, 1902.

Ellacombe, Canon. *In a Gloucestershire Garden.* London: Edward Arnold, 1895.

Ewing, Juliana Horatia. *Dandelion Clocks and Other Tales.* New York: E. & J.B. Young and Co., 1894.

"Fairies in a Surbiton Garden". *The Garden,* August 24, 1918.

Gerard, John. *The Herbal.* New York: Dover Publishing, 1980. First published in 1633.

Gibson, William Hamilton. *Sharp Eyes: A Rambler's Journal.* New York: Harper and Brothers, 1892.

Gordon, Elizabeth. *Flower Children.* New York: Wise and Parslow Co., 1939.

Greenwood, Laura. *The Rural Wreath: Life Among the Flowers.* Boston: Wentworth and Co., 1856.

Hadfield, Miles. *The Gardener's Companion.* London: J.M. Dent and Sons, Ltd., 1936.

Haines, Jennie Day. *De Gardenne Boke.* San Francisco: Paul Elder and Co., 1906.

Haynes, Louise Marshall. *Over the Rainbow Bridge.* Illinois: P.F. Volland Col, 1920.

Howe, Elias. *The Language of Flowers.* New York: Leavitt and Allen, 1847.

Jekyll, Gertrude. *Children and Gardens.* Suffolk, England: Antique Collector Club, 1984. First published in 1908.

Johnson, A.T. "The Country of the Little People". *The Garden,* May 18, 1918.

Kelman, Janet, and Olive Allen. *Gardens Shown to the Children.* New York: Platt and Peck Co., 1899.

Lawrence, Elizabeth. *Gardening for Love.* Durham, North Carolina: Duke University Press, 1987.

Leist, Velista Preston. "Unbuttoning the Peas". *Child Life Magazine,* May, 1937.

Lounsberry, Alice. *The Wildflower Book for Young People.* New York: Frederick A. Stokes Co., 1906.

Miller, Olive Beaupre. *Through the Gate.* Vol. IV, My Book House. Chicago: The Book House for Children, 1920.

Morley, Margaret W. *Flowers and Their Friends.* Boston and New York: Ginn and Co., 1897.

Mulets, Lenore E. *Tree Stories.* Boston: L.C. Page and Co., 1904.

Paine, Albert B. *A Little Garden Calendar.* Philadelphia: Henry Altemus Co., 1905.

Rohde, Eleanour Sinclair. *A Chaplet of Flowers.* London: Medici Society, n.d.

Shafer, Sara Andrew. *A White Paper Garden.* Chicago: A.C. McClurg and Co., 1910.

Stack, Frederick William. *Wildflowers Every Child Should Know.* New York: Doubleday, Page and Co., 1909.

Stevenson, Robert Louis. *A Child's Garden of Verses.* Boston: L.C. Page and Co., 1900.

"Sunflower Competition, The". *The Garden,* October 12, 1918.

Tice, Patricia M. *Gardening in America 1830-1910.* Rochester, New York: The Strong Museum, 1984.

Walker, Margaret Coulson. *Lady Hollyhock and Her Friends.* New York: The Baker and Taylor Co., n.d.

Waterman, Catharine. *Flora's Lexicon, or The Language and Sentiment of Flowers.* Philadelphia: Hooker and Claxton, 1839.

Waugh, Ida. *Holly Berries.* New York: E.P. Dutton Co., 1881.

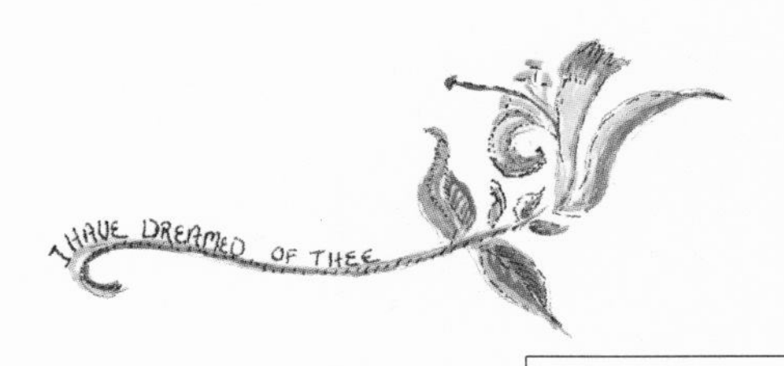

Glossary of Plant Names

African marigold *Tagetes erecta*
Alpine strawberry *Fragaria vesca*
Angel's trumpet *Datura inoxia*
Anise hyssop *Agastache foeniculum*
Bachelor's button *Centaurea cyanus*
Balloon flower *Platycodon grandiflorus*
Beauty-of-the-night (see Four o'clock)
Bee balm *Monarda didyma*
Bishop's hat *Astrophytum myriostigma*
Black eyed susan *Rudbeckia hirta*
Bleeding heart *Dicentra spectabilis*
Box *Buxus sempervirens*
Brampton, queen's stock *Matthiola incana*
Buttercup *Ranunculus*
Cabbage *Brassica oleracea*
Calendula, mary's gold *Calendula officinalis, C. arvensis*
California poppy *Eschscholzia californica*
Campion *Silene alba*
Canterbury bell *Campanula medium*
Cape marigold *Dimorphotheca* spp.
Carnation *Dianthus cariophyllus*
Carrot *Daucus carota*
Catalpa *Catalpa bignoniodes*
Cat's ear *Hypochoeris maculosa*
Cedar *Cedrus* spp.
Chamomile *Chamaemelum nobile*
Chickweed *Stellaria media*
Chicory *Cichorium intybus*
Chinese lantern *Physalis alkekengi*
Clover *Trifolium*
Cockscomb, coxcomb *Celosia cristata*
Colts foot *Galax urceolata*
Columbine *Aquilegia* spp.
Convolvulus *Convolvulus*
Coral bells *Heuchera sanguinea*
Cornflower (see Bachelor's button)
Cosmos *Cosmos bipinnatus*
Cowbell *Uvularia*
Cowslip *Primula veris*
Cress (see Garden cress)
Crocus *Crocus vernus*
Cucumber *Cucumis sativus*
Cup-and-saucer vine *Cobaea scandens*
Daffodil *Narcissus*
Dandelion *Taraxicum officinalis*
Daisy *Erigeron annuus*
Daughter-of-the-evening (see Sweet rocket)
Day lily *Hemerocallis* spp.
(night blooming) *H. altissima*
Delphinium *Delphinium*
Dogwood *Cornus stolonifera*
Dollar plant (see Money plant)
Easter egg eggplant *Solanum melongena*
Egyptian water lily *Nymphaea lotus*
Emilia *Emilia sonchifolia*
English daisy *Bellis perennis*
English wallflower *Cheiranthus cheiri*
Evening lychnis (see Campion)
Evening primrose *Oenothera marginata, O. hookeri*
Evening-scented stock (see Night blooming stock)
Fairy berry (see Alpine strawberry)
False dragonhead *Physostegia virginiana*
Fennel *Foeniculum vulgare*

Field marigold *Calendula arvensis*
Fig marigold *Glottiphylum depressum*
Filaree *Erodium cicutarium*
Flag (see Iris)
Flanders poppy (see Shirley poppy)
Flax *Linum grandifolium*
Forget-me-not *Myosotis sylvatica*
Four o'clock *Mirabilis jalapa*
Foxglove *Digitalis purpurea*
Fraises des bois (see Alpine strawberry)
Fringed pink *Silene laciniata*
Fuchsia *Fuchsia*
Garden cress *Lepidium sativum*
Garden verbena (see Verbena)
Gazania *Gazania rigens*
Geranium *Pelargonium* spp.
Germander *Teucrium* spp.
Gladiolus *Gladiolus*
Goat's beard *Tragopogon pratensis*
Goldenrod *Solidago odora*
Golden star *Bloomeria crocea*
Gooseberry *Ribes uva-crispa*
Goose grass *Eleusine indica*
Gourd *Lagenaria* spp.
Granny's bonnet (see Columbine)
Grape hyacinth *Muscari botryoides*
Hawkbit *Leontodon*
Heart's ease *Viola tricolor*
Heavenly blue morning glory *Ipomoea rubrocaerule*
Hens-and-chicks *Echeveria*
Hepatica *Hepatica americana*
Hollyhock *Alcea rosea*
Honesty (see Money plant)
Honeysuckle *Lonicera japonica*
Hops *Humulus lupulus*
Horehound *Marrubium vulgare*
Horseradish *Armoracia rusticana*
Hyssop (see Anise hyssop)
Ice plant *Mesembryanthemum crystallinum*
Iceland poppy *Papaver nudicale*
Indian corn *Zea mays*
Iris *Iris*
Ivy *Hedera helix*
Jack-go-to-bed-at-noon (see Goat's beard)
Jack-in-the-pulpit *Arisaema triphyllum*
Jacob's ladder *Pedilanthus tithymaloides*
Jewelweed *Impatiens capensis*
Jimson weed *Datura inoxia*
Johnny-jump-up (see Heart's ease)
Jonquil (see Daffodil)
Jupiter's beard *Centranthus ruber*
Kale *Brassica oleracea*
Ladies' tresses *Spiranthes cernua*
Lady's mantle *Alchemilla* spp.
Lady's slipper *Cypripedium acaule*
Lamb's ear *Stachys byzantina*
Lamb's quarters *Chenopodium album*
Lantana *Lantana camara*
Lavender *Lavandula* spp.
Lemon verbena *Aloysia triphylla*
Lilac *Syringa vulgaris*
Little doves (see Columbine)
London pride *Lychnis chalcedonica*
Love-in-a-mist *Nigella damascena*

Madwort *Aurinia saxatilis*
Maidenhair fern *Adiantum*
Maiden pink *Dianthus deltoides*
Mallow *Malva* spp.
Marguerite *Chrysanthemum frutescens*
Marigold *Tagetes* spp.
Marvel of Peru (see Four o'clock)
Mayflower *Cardamine praetensis*
Mint *Mentha* spp.
Miss-go-to-bed-at-noon (see Chicory)
Mistletoe *Phoradendron serotinum*
Money plant *Lunaria annua, L. rediviva*
Moon flower *Ipomoea alba*
Morning glory *Ipomoea purpurea*
Moss rose *Portulaca grandiflora*
Muskmelon *Cucumis melo*
Nasturtium *Tropaeolum majus*
Nicotiana *Nicotiana alata*
Night-blooming cereus *Selenicereus grandiflorus*
Night primrose *Oenothera biennis*
Night-flowering campion *Silene noctiflora*
Night blooming stock *Matthiola longipetala*
Nottingham catchfly *Silene nutans*
Oak *Quercus* spp.
Obedient plant (see False dragonhead)
Onion *Allium fistulosum*
Oriental poppy (see Poppy)
Ox-eye daisy *Chrysanthemum leucanthemum*
Painted lady bean *Phaseolus coccineus*
Pansy *Viola x wittrockiana*
Parsley *Petroselinium crispum*
Passion flower *Passiflora alata*
Peony *Paeonia lactiflora*
Peppergrass (see Garden cress)
Peppermint *Mentha x piperata*
Periwinkle *Vinca minor*
Petunia *Petunia*
Phlox *Phlox*
Pigweed *Chenopodium album*
Pineapple sage *Salvia elegans*
Pink *Dianthus* spp.
Pink clover *Trifolium pratense*
Pink sandwort *Arenaria purpurascens*
Poor-man's-weatherglass (see Scarlet pimpernel)
Popcorn *Zea mays* var.
Poppy *Papaver orientale*
Portulaca (see Moss rose)
Postage stamp plant *Schizopetalon walkeri*
Pumpkin *Cucurbita pepo*
Pussy willow *Salix caprea*
Queen-Anne's-lace *Daucus carota*
Queen-of-the-night (see Night-blooming cereus)
Radish *Raphanus sativus*
Ragged sailors (see Chicory)
Rocket (see Sweet rocket)
Rose *Rosa* spp., *Rosa rugosa*
Rose geranium *Pelargonium graveolens*
Sage *Salvia officinalis*
St. Bernard's lily *Anthericum liliago*
Sand spurry *Spergularia*
Salvia *Salvia* spp.
Saxifrage *Saxifraga*
Scarlet pimpernel *Anagalis arvensis*
Scarlet runner bean *Phaseolus coccineus*

Scented geranium *Pelargonium* spp.
Shepherd's purse *Capsella bursa pastoris*
Shirley poppy *Papaver rhoeas*
Smilax *Smilax*
Snapdragon *Antirrhinum majus*
Snow drop *Galanthus*
Sow thistle *Sonchus oleraceus*
Spaghetti squash *Cucurbita* var.
Spearmint *Mentha spicata*
Speedwell *Veronica officinalis*
Spiderwort *Tradescantia virginiana*
Spiraea *Astilbe japonica*
Spring beauty *Claytonia virginica*
Star-of-Bethlehem *Ornithogalum umbellatum*
Stock *Matthiola incana*
Strawberry *Fragaria x ananassa*
Strawberry popcorn *Zea mays* var.
Sunflower *Helianthus annuus*
Sun rose *Helianthemum*
Sweet alyssum *Lobularia maritima*
Sweet fennel (see Fennel)
Sweet pea *Lathyrus odoratus*
Sweet rocket *Hesperis matronalis*
Sweet violet (see Violet)
Sweet white tobacco *Nicotiana alata*
Sweet william *Dianthus barbatus*
Sycamore *Platanus occidentalis*
Tansy *Tanacetum vulgare*
Tasselflower (see Emilia)
Thrift *Armeria*
Thyme *Thymus* spp.
Tiger lily *Lilium lancifolium*
Tomato *Lycopersicum* spp.
Touch-me-not (see Jewelweed)
Trumpet vine *Campsis radicans, C. grandiflora*
Tulip *Tulipa* spp.
Ursinia *Ursinia*
Verbena *Verbena x hybrida*
Vesper iris *Pardanthopis dichotoma*
Walnut *Juglans nigra*
Water lily *Nymphaea* spp.
Violet *Viola odorata*
Virginia creeper *Parthenocissus quinquefolia*
Wake robin *Trillium*
Wallflower *Cheiranthus cheiri*
White lychnis *Viscaria elegans*
White water lily *Nymphaea odorata*
Woolly lamb's ear *Stachys byzantina*
Wormwood *Artemisia absinthium*
Xeranthemum *Xeranthemum*
Yarrow *Achillea millefolia*
Yellow lark's heels (see Nasturtium)
Yerba buena *Satureja douglasii*
Zinnia *Zinnia elegans*
Zucchini *Cucurbita pepo* var.

Puzzle Answers

Answers to Grandmother's Garden: *1. Jack-in-the pulpit 2. Angel's trumpet 3. Jonquil (John Quill) 4. Bachelor Button 5. Lamb's Quarters or Phlox 6. Flag (Iris) 7. Virginia Creeper 8. Tiger Lily 9. Goldenrod 10. Maidenhair fern 11. Lilac (lie Lock) 12. Four-o'clocks 13. Hops 14. Tulips (two lips) 15. Sun Rose 16. Poppy 17. Cress (Caress) 18. Peony 19. Mayflower 20. Snapdragon 21. Carnation (Car and Nation) 22. Mistletoe 23. Smilax*

Answers to "Love Among the Roses": *1. Black-eyed Susan 2. Sweet William 3. Rambler 4. Four o'clock 5. Lad's love 6. Lady's slippers 7. Ladies tresses 8. Wake robin 9. Spring beauty 10. Marguerite 11. English daisy 12. London pride 13. Thrift 14. Bachelor's buttons 15. Goldenrod 16. Fair maids of France 17. Maiden pink 18. Honesty 19. Bleeding heart 20. Wallflower 21. Jacob's ladder 22. Johnny jump-up 23. Coxcomb (cockscomb) 24. Love-in-a-mist 25. Bishop's hat 26. Eyebright 27. Poppy 28. Blush rose 29. Tulips 30. Speedwell*

Suppliers of Seeds, Plants, and Bugs

Friendly Bugs
Rincon-Vitova
P.O. Box 95
Oakview, California 93002.
(800) 248-2847 or
(805) 643-5407
Write for an informative brochure and price list of all the best bugs.

Seeds Blüm
Idaho City Stage
Boise, Idaho 83706
(208) 342-0858
Vegetables from tiny to giant, and much more. Catalog, $3.00.

Shepherd's Garden Seeds
6116 Highway 9
Felton, California 95018
(408) 335-6910
Children's garden, baby vegetable, edible flower collections. Catalog, $1.00

White Swan Seeds
8030 S.W. Nimbus Avenue
Beaverton, Oregon 97005
Butterfly garden mix; available in many seed stores. Write for the dealer nearest you.